HARCOURT
Science

Harcourt School Publishers

Orlando • Boston • Dallas • Chicago • San Diego

www.harcourtschool.com

Cover Image
This butterfly is a Red Cracker. It is almost completely red on its underside. It is called a cracker because the males make a crackling sound as they fly. The Red Cracker is found in Central and South America.

Printed in the United States of America

ISBN 0-15-315686-4   UNIT A
ISBN 0-15-315687-2   UNIT B
ISBN 0-15-315688-0   UNIT C
ISBN 0-15-315689-9   UNIT D
ISBN 0-15-315690-2   UNIT E
ISBN 0-15-315691-0   UNIT F

2 3 4 5 6 7 8 9 10 032 2000

Authors

Marjorie Slavick Frank
Former Adjunct Faculty Member at Hunter, Brooklyn, and Manhattan Colleges
New York, New York

Robert M. Jones
Professor of Education
University of Houston-Clear Lake
Houston, Texas

Gerald H. Krockover
Professor of Earth and Atmospheric Science Education
School Mathematics and Science Center
Purdue University
West Lafayette, Indiana

Mozell P. Lang
Science Education Consultant
Michigan Department of Education
Lansing, Michigan

Joyce C. McLeod
Visiting Professor
Rollins College
Winter Park, Florida

Carol J. Valenta
Vice President—Education, Exhibits, and Programs
St. Louis Science Center
St. Louis, Missouri

Barry A. Van Deman
Science Program Director
Arlington, Virginia

Senior Editorial Advisor

Napoleon Adebola Bryant, Jr.
Professor Emeritus of Education
Xavier University
Cincinnati, Ohio

Program Advisors

Michael J. Bell
Assistant Professor of Early
    Childhood Education
School of Education
University of Houston-Clear Lake
Houston, Texas

George W. Bright
Professor of Mathematics Education
The University of North Carolina at
    Greensboro
Greensboro, North Carolina

Pansy Cowder
Science Specialist
Tampa, Florida

Nancy Dobbs
Science Specialist, Heflin Elementary
Alief ISD
Houston, Texas

Robert H. Fronk
Head, Science/Mathematics
    Education Department
Florida Institute of Technology
Melbourne, Florida

Gloria R. Guerrero
Education Consultant
Specialist in English as a Second
    Language
San Antonio, Texas

Bernard A. Harris, Jr.
Physician and Former Astronaut
(*STS 55—Space Shuttle Columbia,
STS 63—Space Shuttle Discovery*)
Vice President, SPACEHAB Inc.
Houston, Texas

Lois Harrison-Jones
Education and Management
    Consultant
Dallas, Texas

Linda Levine
Educational Consultant
Orlando, Florida

Bertie Lopez
Curriculum and Support Specialist
Ysleta ISD
El Paso, Texas

Kenneth R. Mechling
Professor of Biology and Science
    Education
Clarion University of Pennsylvania
Clarion, Pennsylvania

Nancy Roser
Professor of Language and Literacy
    Studies
University of Texas, Austin
Austin, Texas

Program Advisor and Activities Writer

Barbara ten Brink
Science Director
Round Rock Independent School
    District
Round Rock, Texas

Reviewers and Contributors

Dorothy J. Finnell
Curriculum Consultant
Houston, Texas

Kathy Harkness
Retired Teacher
Brea, California

Roberta W. Hudgins
Teacher, W. T. Moore Elementary
Tallahassee, Florida

Libby Laughlin
Teacher, North Hill Elementary
Burlington, Iowa

Teresa McMillan
Teacher-in-Residence
University of Houston-Clear Lake
Houston, Texas

Kari A. Miller
Teacher, Dover Elementary
Dover, Pennsylvania

Julie Robinson
Science Specialist, K-5
Ben Franklin Science Academy
Muskogee, Oklahoma

Michael F. Ryan
Educational Technology Specialist
Lake County Schools
Tavares, Florida

Judy Taylor
Teacher, Silvestri Junior High
    School
Las Vegas, Nevada

UNIT A

LIFE SCIENCE
Plants and Animals

UNIT B

LIFE SCIENCE
Plants and Animals Interact

UNIT C

EARTH SCIENCE
Earth's Land

UNIT D

EARTH SCIENCE
Cycles on Earth and In Space

UNIT E

PHYSICAL SCIENCE
Investigating Matter

UNIT F

PHYSICAL SCIENCE
Exploring Energy and Forces

Using Science Process Skills

When scientists try to find an answer to a question or do an experiment, they use thinking tools called process skills. You use many of the process skills whenever you think, listen, read, and write. Think about how these students used process skills to help them answer questions and do experiments.

Maria is interested in birds. She carefully observes the birds she finds. Then she uses her book to identify the birds and learn more about them.

Try This Find something outdoors that you want to learn more about. Use your senses to observe it carefully.

Talk About It What senses does Maria use to observe the birds?

Process Skills

Observe — use your senses to learn about objects and events

Charles finds rocks for a rock collection. He observes the rocks he finds. He compares their colors, shapes, sizes, and textures. He classifies them into groups according to their colors.

Try This Use the skills of comparing and classifying to organize a collection of objects.

Talk About It What other ways can Charles classify the rocks in his collection?

Process Skills

Compare — identify characteristics of things or events to find out how they are alike and different

Classify — group or organize objects or events in categories based on specific characteristics

Katie measures her plants to see how they grow from day to day. Each day after she **measures** she **records the data**. Recording the data will let her work with it later. She **displays the data** in a graph.

Try This Find a shadow in your room. Measure its length each hour. Record your data, and find a way to display it.

Talk About It How does displaying your data help you communicate with others?

Process Skills

Measure — compare mass, length, or capacity of an object to a unit, such as gram, centimeter, or liter

Record Data — write down observations

Display Data — make tables, charts, or graphs

An ad about low-fat potato chips claims that low-fat chips have half the fat of regular potato chips. Tani **plans and conducts an investigation** to test the claim.

Tani labels a paper bag Regular and Low-Fat. He finds two chips of each kind that are the same size, and places them above their labels. He crushes all the chips flat against the bag. He sets the stopwatch for one hour.

Tani **predicts** that regular chips will make larger grease spots on the bag than low-fat chips. When the stopwatch signals, he checks the spots. The spots above the Regular label are larger than the spots above the Low-Fat label. Tani **infers** that the claim is correct.

Try This Plan and conduct an investigation to test claims for a product. Make a prediction, and tell what you infer from the results.

Talk About It Why did Tani test potato chips of the same size?

Process Skills

Plan and conduct investigations— identify and perform the steps necessary to find the answer to a question

Predict— form an idea of an expected outcome based on observations or experience

Infer— use logical reasoning to explain events and make conclusions

You will have many opportunities to practice and apply these and other process skills in *Harcourt Science.* An exciting year of science discoveries lies ahead!

Safety in Science

Here are some safety rules to follow.

1 **Think ahead.** Study the steps and safety symbols of the investigation so you know what to expect. If you have any questions, ask your teacher.

2 **Be neat.** Keep your work area clean. If you have long hair, pull it back so it doesn't get in the way. Roll up long sleeves. If you should spill or break something, or get cut, tell your teacher right away.

3 **Watch your eyes.** Wear safety goggles when told to do so.

4 **Yuck!** Never eat or drink anything during a science activity unless you are told to do so by your teacher.

5 **Don't get shocked.** Be sure that electric cords are in a safe place where you can't trip over them. Don't ever pull a plug out of an outlet by pulling on the cord.

6 **Keep it clean.** Always clean up when you have finished. Put everything away and wash your hands.

In some activities you will see these symbols. They are signs for what you need to do to be safe.

CAUTION

Be especially careful.

CAUTION

Wear safety goggles.

CAUTION

Be careful with sharp objects.

CAUTION

Don't get burned.

CAUTION

Protect your clothes.

CAUTION

Protect your hands with mitts.

CAUTION

Be careful with electricity.

PHYSICAL SCIENCE

Exploring Energy and Forces

Unit Project

Shadow Show

Present a shadow show for another class. Collect a group of toys that move, and write an action story about them. Find some flashlights, and a light-colored sheet. Set up the sheet as a screen. Stand behind it, with the toys between you and the screen. Make shadows by shining the flashlights on the toys as they are moved and the story is read. Ask your audience to describe the forces that are moving the toys in your shadow show.

Heat

Have you ever been to a crowded basketball game? The more the people move around, the hotter it gets. Matter acts in much the same way. Matter is made of little particles that move. The more the particles move, the hotter the matter gets.

Vocabulary Preview

energy
thermal energy
heat
conductor
insulator
thermometer

FAST FACT

Using heat can be lots of fun! Hot-air balloons are able to fly because hot air is pushed up by denser cold air. One of the first hot-air balloons flew in 1783. The first fliers weren't people. They were a sheep, a rooster, and a duck!

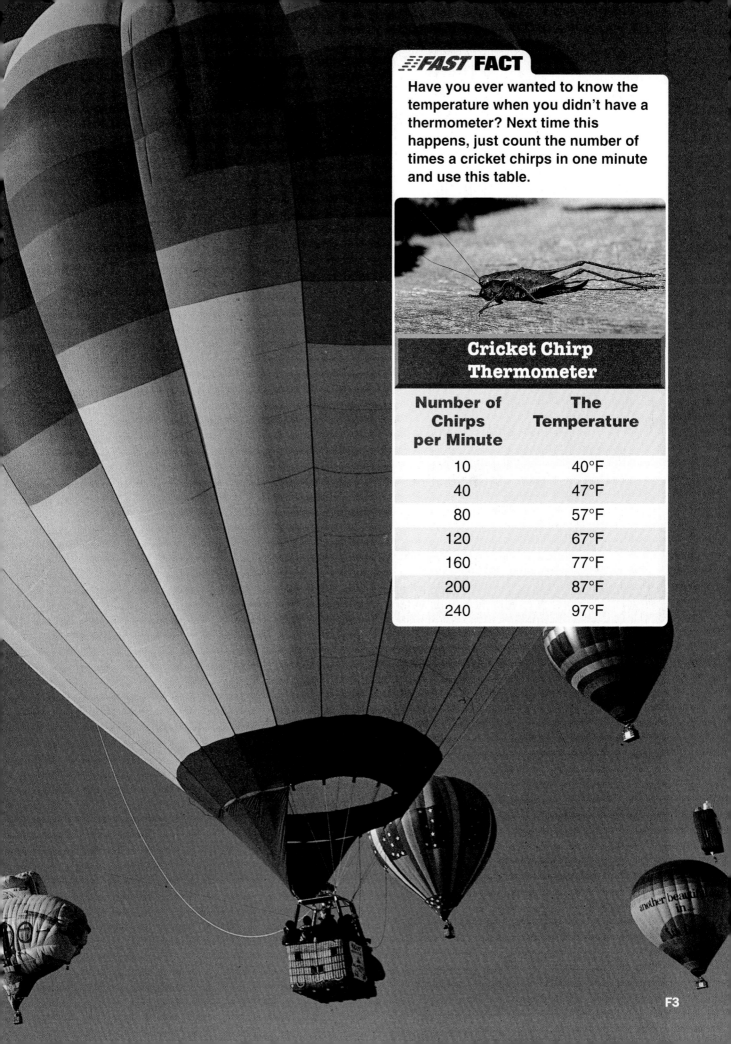

Have you ever wanted to know the temperature when you didn't have a thermometer? Next time this happens, just count the number of times a cricket chirps in one minute and use this table.

Cricket Chirp Thermometer

Number of Chirps per Minute	The Temperature
10	40°F
40	47°F
80	57°F
120	67°F
160	77°F
200	87°F
240	97°F

LESSON 1

What Is Heat?

In this lesson, you can . . .

 INVESTIGATE a way to get thermal energy.

 LEARN ABOUT thermal energy and heat.

 LINK to math, writing, health, and technology.

Rubbing Objects Together

Activity Purpose Scientists use what they know to make statements about how the world works. Then they do tests to see if their statements are correct. The statements they test are called **hypotheses**. In this investigation you will make hypotheses about heat.

Materials
- metal button
- small piece of wool
- penny
- sheet of paper

Activity Procedure

1 Hold your bare hands together, palms touching. Do your hands feel cold, hot, sticky, or damp? **Record** what you **observe**.

◀ Standing close to a grill, you can feel the heat. It begins in the coals. Then the sides of the grill get hot. When the fire is ready for cooking, almost any part of the grill feels hot.

2 Make a **hypothesis**. What would you feel if you rubbed your hands together? **Record** your hypothesis.

3 Now rub your palms together very fast for about ten seconds. **Record** what you **observe**. (Picture A)

Picture A

4 What might you feel if you rubbed the button with the wool? The penny with the paper? Form a **hypothesis** for each test. **Record** each one.

5 Rub the button with the wool for about ten seconds. Touch the button. Touch the wool. **Record** what you **observe**. (Picture B)

6 Then rub the penny with the paper for about ten seconds. Touch the penny. Touch the paper. **Record** what you **observe**.

Picture B

Draw Conclusions

1. What changes did you **observe**? What actions did you perform that caused the changes you observed?

2. Were your **hypotheses** correct? If not, how could you change them based on what you learned?

3. **Scientists at Work** Scientists use their knowledge and experiences to help them **hypothesize**. What knowledge and experiences did you use in this investigation to help you hypothesize?

Process Skill Tip

A **hypothesis** is an *if . . . then* statement. It tells what will happen if something else happens. For example, suppose you noticed that you see bugs only in the summer. You could say "*If* it is summer, *then* there are bugs outside."

Thermal Energy

Energy

FIND OUT

• what thermal energy is

• what heat is

• what thermal energy can do

VOCABULARY

energy
thermal energy
heat

It's hot outside. My soup's cold. Don't touch the stove—it's hot. It's cold out—put on your coat. You've heard these things many times. You know what hot and cold feel like. But do you know what causes them? It's heat.

The story of heat begins with energy. **Energy** is the ability to cause change. There are many ways matter can change. A sheet of paper can be cut into pieces. Baking soda and vinegar can be mixed together to make a new kind of matter. But these changes don't happen by themselves. The paper doesn't cut itself. The baking soda and vinegar don't mix themselves. Energy has to be added to make the changes happen. Your energy cuts the paper. Your energy also mixes the baking soda and the vinegar. Energy in the baking soda and the vinegar makes the new kind of matter. There are many different kinds of energy. All kinds of energy cause changes in matter.

✔ **What is energy?**

◀ The energy in the flame melts the candle wax. This energy makes the matter in the candle change states.

You use energy to fold paper. Your energy changes the paper into a different shape. ▼

Thermal Energy

Even though you can't see them, the particles in all matter are always moving. The particles in solids move back and forth in place. The particles in liquids slide past each other. The particles in gases fly off in all directions. If you have ever moved a heavy chair across the living room, you know it takes a lot of energy. Moving anything from one place to another takes energy. Even moving the tiny particles in matter takes energy. The energy that moves the particles in matter is called **thermal energy**.

We feel thermal energy as heat. Something that is hot has more thermal energy than something that is cold. The particles in a hot stove are moving very fast. They have a lot of thermal energy. The particles in a bowl of cold soup are moving more slowly. They have less thermal energy.

✔ **What is thermal energy?**

A horseshoe is a solid. Its particles are packed tightly together, but they still move a little bit. ▶

A blacksmith puts the horseshoe into the fire. The thermal energy from the fire moves into the horseshoe. When the particles in the fire bump into the particles in the horseshoe, they make them start to move faster too. ▼

Heat

Thermal energy can move from one thing to another. Suppose you warm some cocoa on the stove. Then you pour the cocoa into a cup. When you touch the cup, it feels hot. Thermal energy moves from the hot cocoa to the cool cup. Then thermal energy moves from the warmed cup to your hand. The movement of thermal energy from one place to another is called **heat**.

When you touch a hot cup of cocoa, thermal energy moves from the cup to your hand. If you touch a cold glass of milk, thermal energy moves from your hand to the cold glass. Thermal energy always moves from a hot place to a cold place. It never moves from a cold place to a hot place.

✔ **What is heat?**

These children are blowing warm air onto their hands. Doing this keeps their hands warm on a chilly day. ▼

▲ Fire produces thermal energy. The thermal energy from the stove is used to cook food.

Ways to Make Things Hot

Every time two objects rub together, they produce thermal energy. In the investigation you rubbed your hands together. The particles in your left hand rubbed against the particles in your right hand. The rubbing made the particles in both hands move faster. So you felt heat.

Burning also produces thermal energy. When a gas stove is lighted, the energy stored in the gas particles changes into thermal energy. The thermal energy cooks your food.

Some materials produce thermal energy when they are mixed together. When iron oxide and aluminum are mixed together, they make a hot fire. The fire is so hot that it can even melt metals.

Some gardeners like to use natural materials to make the soil richer. They pile up leaves, soil, food scraps, and grass clippings. Tiny living things in the soil, called *bacteria,* "eat" the leaves and other matter. The bacteria change the mixture into a rich, dark soil-like material called *compost.* During this change, thermal energy is produced.

✔ **Name three ways to produce thermal energy.**

▲ Aluminum mixes with iron oxide to produce a lot of thermal energy.

▲ A gardener turns the compost pile. Turning mixes the bacteria. The more the bacteria "eat," the more thermal energy is produced.

Ways to Use Thermal Energy

You may not think about it, but you use thermal energy every day. It cooks your food. It keeps you warm. You probably use hot air to dry your clothes and hair. You wash your face with warm water. Factories use thermal energy to form metal parts for cars and refrigerators. On the Fourth of July, thermal energy helps give us fireworks. People could not live without thermal energy.

✓ **How do we use thermal energy?**

◀ This artist uses thermal energy to form glass into many shapes, such as this horse.

Factories use thermal energy to melt metal. ▶

▲ A warm home lets you enjoy the snow.

Summary

Energy is the ability to cause matter to change. Thermal energy is the energy that moves the particles inside matter. Thermal energy that moves from one thing to another is called heat.

Review

1. How can you use energy?
2. What does thermal energy do to matter?
3. Name two ways thermal energy can be made.
4. **Critical Thinking** Why do we need thermal energy?
5. **Test Prep** Which is a good definition of heat?
   A ice melting
   B thermal energy that moves
   C fire
   D the energy that moves the particles in matter

LINKS

MATH LINK

Cooking A pie has to bake at 425°F for 15 minutes. Then it has to bake at 350°F for 45 minutes. What is the difference between the two temperatures?

WRITING LINK

Narrative Writing—Story
Write a story about a day without thermal energy. Read your story to a younger child.

HEALTH LINK

Keeping Warm People have used wool for centuries to keep warm. Find out four kinds of animals that we get wool from.

TECHNOLOGY LINK

Learn more about ways thermal energy can be used by watching *Solar Float* on the **Harcourt Science Newsroom Video**.

How Does Thermal Energy Move?

In this lesson, you can . . .

INVESTIGATE what kinds of objects get hot.

LEARN ABOUT three ways thermal energy moves.

LINK to math, writing, health, and technology.

INVESTIGATE

What Gets Hot

Activity Purpose When you were little, you learned not to touch the stove. You had to learn this lesson to keep safe. Now that you are older, you can learn what kinds of things may get hot and what things almost never get hot. **Experiment** to find out.

Materials

- wooden spoon
- plastic spoon
- metal spoon
- 3 foam cups
- water
- ceramic mug
- plastic cup with handle
- metal cup with handle

Activity Procedure

CAUTION

1 Make charts like the ones shown to **record** your **observations**.

2 Touch the three spoons. Are they hot or cold? **Record** what you **observe**.

3 **CAUTION** Be careful with hot water. **It can burn you**. Fill three foam cups with hot tap water. Set them on the table in front of you. Put one spoon in each cup. Wait one minute.

◄ Two colored liquids are inside this lamp. They move because one liquid gets hotter than the other.

Spoons

Wooden	Plastic	Metal

Cups

Ceramic	Plastic	Metal

Picture A

4 Gently touch each spoon. Which one is hottest? **Record** how hot each spoon is. Use words like *cool, warm,* or *hot.* (Picture A)

5 Next, fill each cup with hot tap water. Wait one minute. Then gently touch the handle of each cup. Which one is hottest? (Picture B)

6 Use the words you wrote in Step 4 to **record** how hot each cup handle is.

Picture B

Draw Conclusions

1. Study your Spoons chart. Did all the spoons get hot? Which spoon got the hottest?

2. Look at both charts. Which material got hot in both experiments? How did the plastic items change in the hot water?

3. **Scientists at Work** As they test their ideas, scientists may change their **experiments** in small ways. Suppose you want to see if wood always stays cool when it is placed near heat. What kinds of wooden objects could you use in your experiment?

Process Skill Tip

Scientists wonder how things work. They form hypotheses to explain how things work. To test their hypotheses, they do experiments. An **experiment** is a kind of test done very carefully to find out new information.

The Movement of Thermal Energy

FIND OUT

- three ways thermal energy moves
- how to keep thermal energy from moving easily

VOCABULARY

conductor
insulator

Bumping Particles

In the investigation you put a metal spoon into hot water. At first the spoon was cool. But after it had been in the water, it got hot. Thermal energy moved from the hot water to the cool spoon.

There are three ways thermal energy can move from one place to another. The first way happens √ when things are touching each other.

When objects touch each other, their particles bump into each other. The particles in a hot object move faster than the particles in a cold object. The faster, hotter particles give some of their energy to the slower, cooler particles. Then the slower particles speed up and get hotter.

✔ **How does thermal energy move between objects that touch each other?**

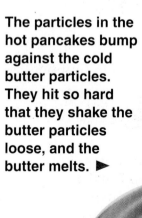

The particles in the hot pancakes bump against the cold butter particles. They hit so hard that they shake the butter particles loose, and the butter melts. ▶

Conductors

Thermal energy moves very quickly in some materials. Metals are an example. Most cooking pots and pans are made of metal because these materials let thermal energy move easily. The energy has to move from the hot stove to the pan to the food. Then the thermal energy will cook the food. Materials in which thermal energy moves easily are called **conductors**.

✔ **What is a conductor?**

▲ The iron in the Dutch oven makes this a good cooking pot. The iron heats easily so the food can cook.

Insulators

Sometimes you don't want thermal energy to move from one object to another. Suppose you are using a metal cookie sheet to use the thermal energy in the oven to bake some cookies. But then you need to take the cookie sheet out of the oven. Now you need something to keep the thermal energy away from your hands! You need an insulator. An **insulator** is a material in which thermal energy can't move easily. Its particles don't move much when particles from a hot object bump into them. Oven mitts are good insulators.

✔ **What is an insulator?**

▲ This pizza holder does not take in thermal energy easily. The thermal energy from the pizza moves very slowly to the holder. The holder is an insulator.

Moving Liquids and Gases

The second way thermal energy moves from one place to another is in moving liquids and gases. The particles in liquids and gases can move from one place to another. Particles in hot liquids and gases often move to new places before they transfer thermal energy. In some homes, furnaces use fire to heat air. Then fans blow the hot air into the cold rooms in the house. The hot air from the furnace gives its thermal energy to the whole house.

Even when they're not pushed by fans, hot liquids and gases move. When a liquid or gas is heated, its particles spread out. This makes the hot liquid or hot gas lighter. Hot air is lighter than cold air. Cold air sinks to the floor, pushing the hot air up.

✔ **Why does hot air move up?**

▲ **In the desert the hot ground warms the air. The hot air is pushed up by cooler air. As the hot air moves, it makes things around it look wavy.**

Heat from the Sun

The third way thermal energy moves from one place to another happens when things aren't touching each other. When you feel warm standing in the sun, you are not touching the sun. So how do you get warm without touching something that is warm? Thermal energy that moves without touching anything is called *radiation*.

✔ **How can thermal energy move without touching anything?**

◄ **Energy from the sun warms the air and the Earth. We cannot touch the sun, but we can feel its heat.**

Animals and humans depend on the sun to keep warm. ▶

Summary

When things touch each other, their particles bump and thermal energy moves between them. Conductors let thermal energy travel easily. Insulators are materials in which thermal energy doesn't move easily. Heated liquids and gases can move before they transfer their thermal energy. We get thermal energy from the sun without touching it.

Review

1. Describe the three ways thermal energy can move.
2. Is a potholder a conductor or an insulator? Explain.
3. Is the metal in a pot a conductor or an insulator? Explain.
4. **Critical Thinking** Animals that have wool keep warm because of their wool. A sweater or coat made of wool keeps you warm too. Why?
5. **Test Prep** Which is the best way to describe what happens to air when it gets warm?

   A Warm air sinks.

   B The particles in warm air get close to each other.

   C Warm air is pushed up by cooler air.

   D Warm air is very still.

LINKS

MATH LINK

A Problem Suppose the burner under a pot of water is turned on at 2:15 P.M. It begins to boil at 2:35 P.M. How long did it take the water to come to a boil?

WRITING LINK

Informative Writing— Explanation Some sports are often done in the warm sunshine. Write a paragraph for your teacher explaining why people who swim outdoors usually do so only when it's warm.

HEALTH LINK

Safety Rules People need thermal energy, but using it can be dangerous. Write two rules for staying safe in the kitchen when food is cooking on the stove.

TECHNOLOGY LINK

Visit the Harcourt Learning Site for related links, activities, and resources.

www.harcourtschool.com

WELCOME TO THE LEARNING SITE

How Is Temperature Measured?

In this lesson, you can . . .

INVESTIGATE tools for measuring temperature.

LEARN ABOUT ways to control thermal energy.

LINK to math, writing, social studies, and technology.

INVESTIGATE

Measuring Temperature

Activity Purpose Many people don't leave home in the morning without finding out what the temperature outside is. Knowing the temperature is especially handy for getting dressed! In this investigation you will find out how to **measure** temperature.

Materials
- 2 cups
- water
- thermometer

Activity Procedure

1 Make a chart like the one shown.

2 **CAUTION** **Be careful with hot water. It can burn you.** Fill one of the cups with cold tap water. Fill the other with hot tap water.

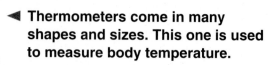

◀ **Thermometers come in many shapes and sizes. This one is used to measure body temperature.**

3 Put the thermometer into the cup of cold water. **Observe** what happens to the liquid in the thermometer. On your chart, **record** the temperature. (Picture A)

Water	Temperature
Cold water, Step 3	
Hot water, Step 4	
Cold water, Step 5	
Hot water, Step 6	

4 Put the thermometer into the cup of hot water. **Observe** what happens to the liquid in the thermometer. On your chart, **record** the temperature.

5 Put the thermometer back into the cup of cold water. **Observe** what happens to the liquid in the thermometer. On your chart, **record** the temperature.

Picture A

6 Wait one minute. Put the thermometer back into the cup of hot water. **Observe** what happens to the liquid in the thermometer. On your chart, **record** the temperature.

Draw Conclusions

1. What happened to the liquid inside the thermometer each time you put the thermometer in the hot water? What happened when you put it in the cold water?

2. Was the temperature of the hot water the same both times you **measured** it? What do you think caused you to get the measurements you got?

3. **Scientists at Work** Scientists **measure** carefully. Write directions for how to measure temperature by using a thermometer.

Process Skill Tip

Scientists look for answers to questions, but they do not ever make up the answers. To try to answer their questions, they **measure** with care. Then anyone who wants to check the answers can take the same measurements.

Measuring Temperature

FIND OUT

- how to measure temperature
- ways to control thermal energy

VOCABULARY

thermometer

Thermometers

In the investigation you used a thermometer. A **thermometer** is a tool used to measure how hot or cold something is. The liquid in your thermometer expanded when it got hot. It needed more space, so it moved up the tube. Many thermometers work this way. When the liquid cools, its particles don't move as fast. They get closer together. Then the liquid sinks down the tube.

Thermometers have numbers printed on them much like rulers do. Instead of inches or centimeters, they use scales called Fahrenheit and Celsius. By learning to read these scales, you can read the temperature from a thermometer.

✔ **How does a liquid thermometer work?**

▲ One kind of thermometer has the scale printed in a circle. It is a dial thermometer. It uses a metal coil instead of a liquid.

Scientists have measured the temperatures of many kinds of things. Some examples are shown here.

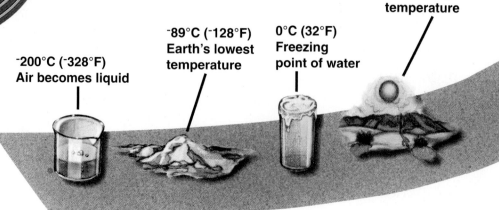

⁻200°C (⁻328°F) Air becomes liquid

⁻89°C (⁻128°F) Earth's lowest temperature

0°C (32°F) Freezing point of water

58°C (136°F) Earth's highest temperature

F20

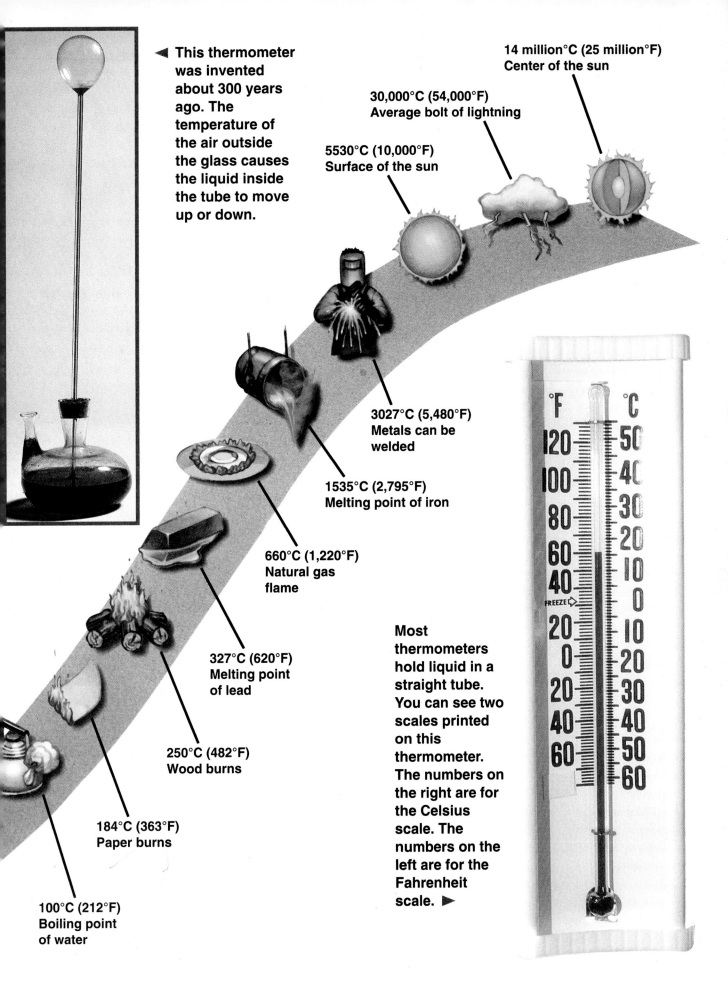

◀ This thermometer was invented about 300 years ago. The temperature of the air outside the glass causes the liquid inside the tube to move up or down.

14 million°C (25 million°F)
Center of the sun

30,000°C (54,000°F)
Average bolt of lightning

5530°C (10,000°F)
Surface of the sun

3027°C (5,480°F)
Metals can be welded

1535°C (2,795°F)
Melting point of iron

660°C (1,220°F)
Natural gas flame

327°C (620°F)
Melting point of lead

250°C (482°F)
Wood burns

184°C (363°F)
Paper burns

100°C (212°F)
Boiling point of water

Most thermometers hold liquid in a straight tube. You can see two scales printed on this thermometer. The numbers on the right are for the Celsius scale. The numbers on the left are for the Fahrenheit scale. ▶

Controlling Thermal Energy

Humans can die from being too cold or too hot. People's bodies need to stay at a certain temperature. We can live in cold weather and in hot weather because we have learned how to control heat.

One way we control heat is by wearing clothes. In hot weather we dress lightly. In cold weather we dress warmly. Animals adjust to different temperatures also. In the winter a dog's coat gets thicker. By summer its fur is thinner.

One way to control heat inside a building is to use a thermostat. Almost every building has one. It looks like a little box or circle on the wall. Thermostats have temperature scales, just like thermometers. In fact, they have thermometers inside them.

✓ **Name two ways people control heat.**

THE INSIDE STORY

Thermostats

A thermostat turns the furnace and the air conditioner off and on. Most large buildings, like schools, have several furnaces. Each furnace or air conditioner has its own thermostat.

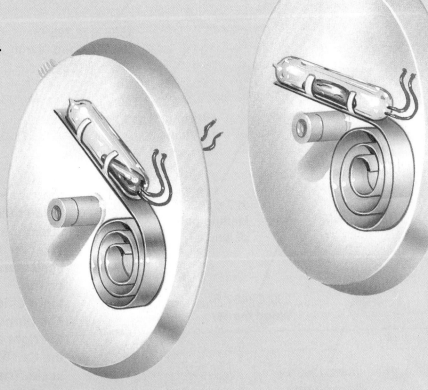

Inside this thermostat is a coiled piece of metal made of two different kinds of metals glued together. Each kind of metal gets bigger when it gets warm. But the metal on the bottom gets bigger faster than the metal on the top. So when the room gets warmer, the coil gets bigger. It turns on the switch for the air conditioner. Then the air conditioner can cool the room.

◄ This thermostat is round. You can set it to the temperature you want in the room.

Summary

Thermometers are used to measure temperature. In a liquid thermometer, thermal energy makes the liquid inside the tube move up. We can read the temperature on a scale next to the tube. A thermostat helps control the temperature inside a building.

Review

1. What does a thermometer do?
2. Why does a liquid thermometer work as it does?
3. What kinds of scales are used on thermometers?
4. **Critical Thinking** Why would you need to know what the temperature outside is?
5. **Test Prep** Two different kinds of metals make up thermostat coils because —

   A it's cheaper to make them using two metals

   B they look better that way

   C there is not enough of one kind of metal

   D one metal gets bigger faster than the other

LINKS

MATH LINK

Skip-Counting Look at the scale on a thermometer. Do you need to skip-count by 5s or by 10s to read it?

WRITING LINK

Informative Writing— Description Write a weather forecast for your class. Pretend you are on the TV news. Describe what the weather will be like for the next five days.

SOCIAL STUDIES LINK

Geography Use a world map or a globe to find two countries that have extremely cold winters. Find two tropical countries that are warm all year. Explain how you chose your countries.

TECHNOLOGY LINK

Learn more about measuring temperature by visiting this Internet Site.
www.scilinks.org/harcourt

SCiLINKS
THE WORLD'S A CLICK AWAY

Technology Delivers Hot Pizza

You know that thermal energy always moves from hot objects to cold objects. This means thermal energy moves from a hot pizza to the cooler air around it. In the process, the hot pizza gets cold. And who wants cold pizza?

We Want It Hot

A recent survey found that more than half of all take-out food was delivered to homes rather than picked up. Keeping pizza hot while the delivery person finds the right house is sometimes a problem. Customers want their food delivered fast, and they want it to arrive piping hot. Companies that deliver cold pizzas are soon out of business. So it's good news that science is helping pizza companies deliver your pizza hot to your door.

Thermal Retention

A new product is being used to keep take-out food hot. It is called the hot bag, and it keeps the heat of the food inside the container.

The hot bag is made with three layers of material. The inside liner, made of nylon and vinyl, reduces condensation. It also has a shiny silver color to reflect the heat of the food back inside the container.

The middle layer is a dense foam. This layer is the insulator for the bag. Thermal energy doesn't move easily through the foam. That means the heat is trapped inside the bag. The foam also prevents air from moving into or out of the bag. Even though the inside of the bag is hot, the outside of the bag remains cool and easy to handle.

The outside of the bag is made of heavy vinyl. This makes the bag waterproof. The hot-bag makers use strips of Velcro to seal the bag, keeping even more thermal energy in. The hot bag also has two small openings that allow steam to escape.

They keep the pizza crust from becoming soggy on its trip to your home.

Hot Delivery

The makers of the hot bag promise that food in this bag will remain at 80°C (176°F) for at least 15 minutes. After 30 minutes the temperature inside the bag may drop 5 to 8°C (about 10 to 15°F). But thanks to the new materials, the pizza remains 17°C (31°F) hotter than it would using the old way. The crust stays crisp, and the cheese stays gooey. Hot-bag manufacturers are always watching for new technology that can help them deliver your pizza just the way you like it—hot from the oven.

Think About It

1. What other types of fast-food companies could use the hot bag to help them in their deliveries?

2. Would these bags be helpful for delivering ice cream? Explain.

WEB LINK:
For Science and Technology updates, visit the Harcourt Internet site.
www.harcourtschool.com

foam

pizza

What They Do

Pizza chefs work with fresh ingredients and pizza dough to make pizza. Pizza chefs often work until late at night to make pizzas for parties and other events.

Education and Training A person who is hired to be a pizza chef will spend some time learning on the job from an experienced chef. After training, the new chef is able to work on his or her own.

Percy Spencer

INVENTOR

Have you ever made popcorn in a microwave oven? That's how microwave ovens were invented in 1945. Percy Spencer was touring one of the laboratories at the Raytheon Company, where he worked. He was standing close to the power tube that drives a radar set. Suddenly he realized that the chocolate bar in his pocket was beginning to melt. He asked for some unpopped popcorn to test his theory that the machine could cook food.

Percy Spencer was awarded 120 patents during his lifetime. His patent for the microwave oven was given in 1952. The first microwave oven was more than 5 feet tall and weighed 750 pounds! It was designed for cooking large amounts of food in restaurants, on ocean liners, and on railroad cars.

Today microwaves are used to do many things besides heat food. Microwaves are used in radar. Some researchers are working to use microwaves to sterilize food. Some industries have begun using microwaves as well. Microwaves are used to bake the tiles that protect a space shuttle from heat when it returns to the Earth's atmosphere.

One of the most exciting possibilities is that microwaves might be used to treat some kinds of cancer. Cancer tissue begins to die at 109°F. Researchers hope they can use microwaves to kill the cancer and not harm the healthy tissue.

Think About It

1. How do you know that Percy Spencer was creative?
2. Have you ever had an unexpected discovery? What was it?

Testing Insulators

Is wool or sand a better insulator?

Materials

- 1 small coffee can with plastic lid
- 2 wool socks
- scissors
- 1 long lab thermometer
- clock with a second hand
- hair dryer
- 3 cups of sand

Procedure

1. Stuff the can with the socks. Put the lid on. Cut a small hole in the middle of the lid for the thermometer. Make sure the thermometer is surrounded by wool. After five minutes, read and record the temperature.

2. Start the hair dryer. Warm the outside of the can. Watch the thermometer and the clock. See how long it takes the temperature to go up 10°C.

3. Repeat using sand in the coffee can.

Draw Conclusions

Compare results. Which is the better insulator? How do you know?

Cooling Water

How long does it take for warm water to cool?

Materials

- 2 foam cups
- measuring cup
- 2 thermometers
- clock with a second hand
- warm water

Procedure

1. Fill one cup with 1 cup of warm water. Fill the other cup with $\frac{1}{2}$ cup of warm water.

2. In which cup will the water cool faster? Make a hypothesis to answer the question. Write down your hypothesis.

3. Make a chart to record temperatures. Record the starting temperature in each cup. Then record the temperature of each cup every minute until one cup of water reaches room temperature.

Draw Conclusions

Was your hypothesis correct? Explain.

Chapter ① Review and Test Preparation

Vocabulary Review

Choose a term below to match each definition. The page numbers in ( ) tell you where to look in the chapter if you need help.

energy (F6)
thermal energy (F7)
heat (F8)
conductor (F15)
insulator (F15)
thermometer (F20)

1. The ability to cause change

2. Matter in which thermal energy doesn't move easily

3. The movement of thermal energy from one place to another

4. Matter in which thermal energy moves easily

5. A tool that measures temperature

6. The energy of the moving particles inside matter

Connect Concepts

Write the terms to complete the concept map.

hot   the sun   liquids   gases
particles   thermal energy   cold

Heat is the movement of 7. _____.

Thermal energy always moves from 8. _____ to 9. _____.

HEAT

Two objects are next to each other. Thermal energy moves between them when their 10. _____ bump into each other.

Hot 11. _____ and _____ can move before they give away their thermal energy.

The kind of thermal energy that comes from 12. _____ can move without touching anything.

Check Understanding

Write the letter of the best choice.

13. Which is an insulator?

  A silver spoon

  B aluminum foil

  C stainless steel spoon

  D wooden spoon

14. Which is a conductor?

  F wooden spoon

  G oven mitt

  H aluminum foil

  J wool sweater

15. Heat from the sun gets to Earth by —

  A moving through liquids

  B moving through gases

  C moving without touching anything

  D bumping into Earth's particles

16. Moving anything from one place to another takes —

  F heat   **H** a thermometer

  G energy   **J** a thermostat

Critical Thinking

17. You take a dish out of a hot oven. Before you set the dish on the table, you put down a thick straw pad. Why?

18. You are holding an ice cube. Is thermal energy moving from your hand to the ice or from the ice to your hand? Explain.

Process Skills Review

19. Two people **measure** the temperature of the same cup of hot water. They use two different thermometers. One person says the temperature of the water is 40°C. The other says it is 50°C. What can cause the different answers?

20. What **experiment** could you do to find out why the temperatures in Question 19 were different?

21. It is July. The sun is shining. What **hypothesis** can you state about the temperature? How could you check your hypothesis?

Performance Assessment

Diagram a Thermometer

Water boils at 100°C and freezes at 0°C. Draw a picture of a liquid thermometer in a pot of boiling water. Draw a picture of the same thermometer in a pot of freezing water. Explain what happens to the thermometer when it moves from the boiling water to the freezing water.

Vocabulary Preview

reflection
refraction
absorption
prism

Light

Flick! Bounce, reflect, bounce!
That's what happens to the light from
a flashlight if you turn it on and shine
it at your image in a mirror. The light
goes so fast it seems to hit the mirror
and you at the same instant you turn
the flashlight on!

FAST FACT

We see stars as they were when their light left them. This
table shows how long it takes the light from some space
objects to reach us.

The Speed of Light		
Object in Space	Distance from Earth	Light Reaches Us In
Moon	384,462 km	$1\frac{1}{3}$ seconds
Venus	41.2 million km	$2\frac{1}{3}$ minutes
Sun	149.7 million km	$8\frac{1}{2}$ minutes
Alpha Centauri	40.2 trillion km	$4\frac{1}{3}$ years
Sirius	81.7 trillion km	$8\frac{1}{2}$ years
Andromeda Galaxy	21.2 billion billion km	$2\frac{1}{4}$ million years

Andromeda Galaxy

How Does Light Behave?

In this lesson, you can . . .

INVESTIGATE how light travels.

LEARN ABOUT things light can do.

LINK to math, writing, drama, and technology.

INVESTIGATE

How Light Travels

Activity Purpose You can make shadows with your hands because of the way light travels. In this investigation you will **observe** how light travels.

Materials
- 3 index cards
- ruler
- pencil
- clay
- small, short lamp without a lampshade

Activity Procedure

1 Make a large X on each card. To draw each line, lay the ruler from one corner of the card to the opposite corner. (Picture A)

▲ **Shadow puppets can be fun.**

2 On each card, make a hole at the place where the lines of the X cross. Use the pencil to make the holes.

3 Use the clay to make a stand for each card. Make sure the holes in the cards are the same height. (Picture B)

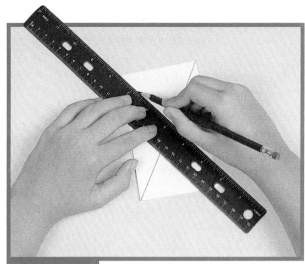

Picture A

4 Turn on the light. Look through the holes in the cards. Move the cards around on the table until you can see the light bulb through all three cards at once. Draw a picture showing where the light is and where the cards are.

5 Move the cards around to new places on the table. Each time you move the cards, draw a picture showing where the cards are. Do not move the light! **Observe** the light through the holes each time.

Picture B

Draw Conclusions

1. Where were the cards when you were able to see the light?

2. Were there times you couldn't see the light? Where were the cards then?

3. **Scientists at Work** Scientists **observe** carefully and then **record** what they observe. Often they draw pictures to **communicate** what they observe. Did drawing pictures help you describe what you saw? Explain.

Light

Light Energy

You know that energy is the ability to cause things to change. The energy in a fire changes a sheet of paper into ashes. The heat from a fire can change your hands from cold to warm. Bacteria use energy to change a dead log into soil for plants.

Light is also a kind of energy. Light energy can make many changes. Without light energy, you could not see anything. Light energy gives things colors. The sun shines on the soil, and plants grow. Light energy can make cars move. In space, light energy powers satellites and space stations. Doctors use the light energy of lasers to perform some operations.

✔ **What are three changes light energy can cause?**

◄ Plants can't live without light. Plants use the sun's light to make food.

▲ Scientists are finding new ways to use the sun's energy. Some new cars use light energy instead of gas.

◀ **The sun provides energy to Earth.**

Shadows

When you put your hand in front of a lamp, you make a shadow on the wall. The shadows move and change shape as you move your hand. Shadows move and change because of the way light travels.

Light travels in straight lines. When you put your hand in front of a lamp, some of the straight lines of light hit your hand. The shadow on the wall shows where the light is blocked by your hand. When you move your hand, the shadow moves because your hand blocks different lines of light.

In the investigation you could see the light bulb only when the holes in the three cards were in a straight line. When one of the holes wasn't in line with the others, it blocked the line of light. How did this show that light travels in a straight line?

When you stand in the sun, you block some of the lines of sunlight. As the sun moves in the sky, you block different lines of light. When the sun is low in the sky, in the morning and in the afternoon, your shadow is long. When the sun is high overhead, your shadow is short.

✔ **How does light travel?**

▲ Shadows caused by sunlight are long in the morning. These shadows always point away from the sun.

▲ In the afternoon the sun is in a different place. Now the shadow points another way, but it still points away from the sun.

F35

Bouncing Light

Look in a mirror. What do you see? You probably see yourself and some of the things around you. You are looking in front of you at the mirror. But the things you see in the mirror are next to you or even behind you. How is this possible?

Hold a lamp in front of a mirror, and you will see the lamp in the mirror. The light from the lamp moves in a straight line to the mirror. When it hits the mirror, it bounces off. It is still traveling in a straight line. But now it's going in a new direction. It is coming straight back to you. The bouncing of light off an object is called **reflection** (rih•FLEK•shuhn). You see objects in a mirror because their light is reflected straight back to you.

◀ When light bounces off a mirror, the light changes direction. The letters on the sign are backward. This is because a mirror reverses an image from left to right.

Light travels in straight lines. Even if it bounces off many mirrors, you can still see the object. If the mirrors are lined up exactly right, you can see many reflections of the object. ▼

Light bouncing off a smooth surface gives an image you can see. A mirror is very smooth. So are shiny metal and still water. You can see yourself in these things. But most things aren't as smooth as mirrors.

Most things are bumpy. When light hits a bumpy surface, each straight line of light goes off in a different direction. Then you don't see any image.

✔ **What is reflection?**

◀ If the water is rippling, each wave reflects light in a different direction. Since the light is traveling in so many directions, it is hard to see a clear picture on the surface of the water.

The water on the lake is so still that it acts like a mirror. ▼

Bending Light

Light doesn't bounce off every surface. There are some things light goes through. That's why you can see through air, water, and glass.

Light travels at different speeds in air, water, and glass. So when light goes from one thing to another, such as from air to glass, it changes speed. Any time light goes from one kind of matter to another, it changes speed. If light hits the new matter straight on, it keeps going straight. But if light hits the new matter at a slant, the light bends. The bending of light when it moves from one kind of matter to another is called **refraction** (rih•FRAK•shuhn).

Light moving from air to glass is like a skater moving from a sidewalk to the grass. If the skater is going straight into the grass, both front wheels hit the grass at the same time. The skater slows down because grass is softer than concrete. But he or she continues to go straight. If the skater does not go straight into the grass, one wheel hits the grass first. The other is still on the sidewalk. The wheel that hits the grass first slows down first. This makes the skater change direction.

✔ **What is refraction?**

Half of this toy diver is in the water. You see the bottom half through the water. The light bends when it hits the water. You see the top half through air. This light isn't bending. So the toy diver looks as if it is broken in two. ▼

Light travels through air and glass. This light hits the glass straight on and keeps going straight. ▶

Here the light hits the glass at an angle. This time the light bends and changes direction. ▼

Here the light is refracted three times. So the pencil looks as if it is broken into four pieces. ▼

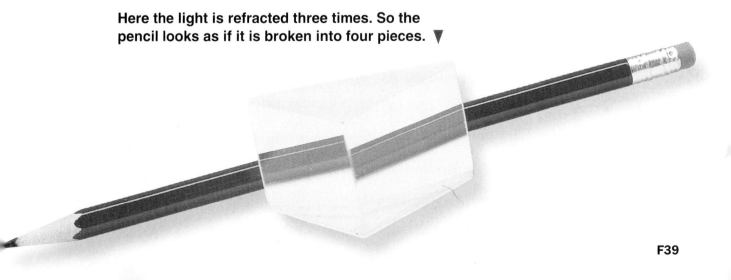

Stopping Light

You have learned that you can see through air, water, and glass. Light travels through these forms of matter. But most matter doesn't let light pass. When light hits a wall, the wall stops, or absorbs, the light. Stopping light is called **absorption** (ab•SAWRP•shuhn). Have you ever watched rain falling on grass? The soil absorbs the water. Most matter absorbs light in the same way.

When light hits most objects, some of the light bounces off and the rest is absorbed. Smooth, shiny objects reflect almost all the light that hits them. Other objects absorb most of the light that hits them and reflect the rest. If an object doesn't produce its own light, what you see when you look at it is the light that bounces off it.

✔ **What is absorption?**

▲ Light travels through this glass window because it is *transparent*. You can see a clear image of the girl through the window. Light also travels through the thin curtain. Matter that lets only some light through is called *translucent*. Light can't travel through the dark curtains. A material that doesn't let light through is *opaque*.

Summary

Light energy can cause things to change. Light travels in a straight line unless it bumps into something. An object that stops light can cause a shadow. Some objects let light pass through them. When light hits an object, it can be reflected, refracted, or absorbed.

Review

1. What does a mirror do?
2. At about what time of day is your shadow shortest?
3. What word describes stopping light so that it is not reflected or refracted?
4. **Critical Thinking** You stick your hand into an aquarium to get something out. Why does your hand look as if it is cut off from your arm?
5. **Test Prep** Which is an example of light energy being used?
   A water boiling
   B a seed sprouting
   C a ball bouncing
   D a girl lifting a chair

LINKS

MATH LINK

Elapsed Time Suppose the sun rises at 6:15 A.M. and sets at 7:15 P.M. How many hours of daylight are there?

WRITING LINK

Informative Writing— Description Write a short story for your classmates that describes a building reflected in a puddle. Include one description for when the water is smooth and one for when the water is rippling.

DRAMA LINK

Shadow Puppets Make a screen out of a cloth sheet. Shine a light behind it. Make shadow animals on the sheet. Use them to tell a story to the class.

TECHNOLOGY LINK

Learn more about how light can be used by watching *Using Natural Light* on the **Harcourt Science Newsroom Video.**

How Are Light and Color Related?

In this lesson, you can . . .

 INVESTIGATE rainbows.

 LEARN ABOUT light and color.

 LINK to math, writing, art, and technology.

INVESTIGATE

Making a Rainbow

Activity Purpose The world is a colorful place. You know that you can see colors only when the light is shining. In the dark you can't see color. So is color in the objects or in the light? In this investigation you can **observe** where colors come from.

Materials

- small mirror
- clear glass
- water
- flashlight

Activity Procedure

1 Gently place the mirror into the glass. Slant it up against the side.

2 Fill the glass with water. (Picture A)

3 Set the glass on a table. Turn out the lights. Make the room as dark as possible.

◀ You can get all the colors of the rainbow in a box of colored pencils.

Picture A

Picture B

4. Shine the flashlight into the glass of water. Aim for the mirror. Adjust your aim until the light hits the mirror. If necessary, adjust the mirror in the water. Make sure the mirror is slanted.

5. **Observe** what happens to the light in the glass. Look at the light where it hits the ceiling or the wall. **Record** what you observe. (Picture B)

Draw Conclusions

1. What did the light look like as it went into the glass?

2. What did the light look like after it came out of the glass?

3. **Scientists at Work** Scientists **draw conclusions** based on what they **observe**. What conclusions can you draw about where color comes from?

Investigate Further Change the angles of the mirror and the flashlight. Which setup gives the best result? Draw a picture of the best arrangement.

Process Skill Tip

You **draw conclusions** when you have gathered data by observing, measuring, and using numbers. Conclusions tell what you have learned.

Light and Color

Prisms

FIND OUT

- **how many colors are in light**
- **what makes a rainbow**

VOCABULARY

prism

Have you ever drawn a picture of the sun? Did you color it yellow? People often do. But sunlight is really made of many different colors. Yellow is only one of them. The sunlight you see is really white light. White is the color of all the sun's colors mixed together.

Different colors of light travel at different speeds in water and in glass. So when white light moves from air to glass or from air to water, the different colors of light bend at different angles. They separate into each individual color.

In the investigation you used water and a mirror to break white light into different colors. Scientists use glass triangle prisms to experiment with light. A **prism** (PRIZ•uhm) is a solid object that bends light. When white light hits the prism, each color of light bends at a different angle. Light that passes through a prism separates into a rainbow.

✔ **What is a prism?**

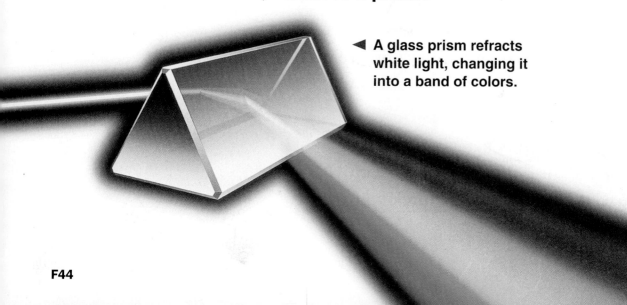

◀ **A glass prism refracts white light, changing it into a band of colors.**

How Rainbows Form

You can sometimes see a rainbow in the sky during a summer rain when the sun is out. ▶

Each drop of falling water is like a tiny prism. So sunlight passes through all the raindrops, and the light rays bend. This bending separates the light into its colors, and you see a rainbow.

When white light separates, its colors always appear in the same order. The order is red, orange, yellow, green, blue, and violet.

Adding Colors

A prism breaks white light into colors. You can also add colors together. When you add different colored lights together, they form other colors. Shining a red light and a green light onto the same spot will make a yellow light. Shining a blue light and a red light onto the same spot will make a purple light. You can add red light, blue light, and green light in different ways to make all other colors.

✔ **What is one method for making colors?**

Seeing Colors

All the colors of light, called white light, hit every object you see. Most objects absorb most of the light, but not all of it. The light that is not absorbed is reflected and is the color you see. For example, green grass absorbs all of the white light except the green part. The green part reflects back to your eyes, and you see green grass.

✔ **Why do you see color?**

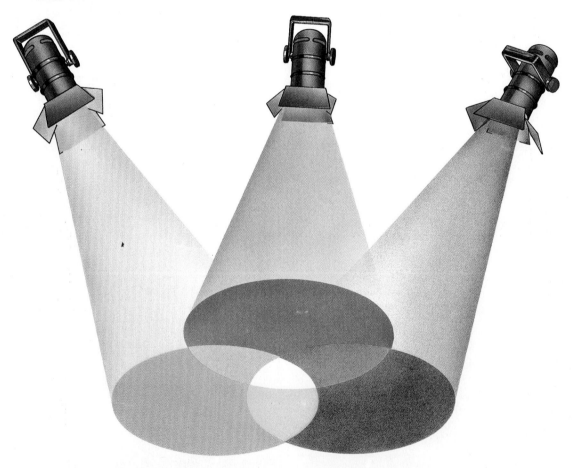

Three basic light colors are red, blue, and green. They will form all other colors. Adding all three of these colors will give white light.

◀ The red rose absorbs all parts of white light except red. Red light is reflected, and we see a red flower.

Summary

White light is made up of many colors mixed together. A prism separates the colors. Raindrops act like prisms to form rainbows. You can make colors by adding different colored lights. The colors of objects you see are the colors of light that the objects reflect.

Review

1. Describe how a prism works.
2. Name the colors that make up white light.
3. What happens if you add different colors of light?
4. **Critical Thinking** Why don't you see a rainbow during most rainstorms?
5. **Test Prep** Which light colors are absorbed by a yellow tulip?
   A red, orange, and yellow
   B red, orange, green, blue, and violet
   C violet, orange, green, and yellow
   D yellow, red, blue, and green

LINKS

MATH LINK

Solid Figures The bases of a triangular prism are triangles. What are the bases of a rectangular prism?

WRITING LINK

Informative Writing— Narration Find five different words that describe colors of red. Write a paragraph for your teacher describing a scene that includes each of these colors.

ART LINK

Color Wheel Find out what a color wheel is and how an artist might use one. Draw one, and explain it to a classmate.

TECHNOLOGY LINK

Visit the Harcourt Learning Site for related links, activities, and resources.
www.harcourtschool.com

WELCOME TO
THE LEARNING SITE

DISCOVERING LIGHT AND OPTICS

We use our eyes to see. A curved lens inside the eye bends light, focusing an image on the retina. This image is sent to the brain, which interprets the image.

Using Lenses

Lenses in tools such as microscopes, telescopes, and even eyeglasses work the same way. All lenses have at least one curved surface. The curve of the lens bends and focuses the light. The image formed by the lens might be smaller than, larger than, or the same size as the original object.

People have worn eyeglasses for hundreds of years. The Italian explorer Marco Polo saw people in China wearing glasses around 1275. After books became common in the late 1400s, glasses became common for reading. During the 1600s, people discovered that using lenses would correct nearsightedness. Nearsighted people have difficulty seeing objects far away. More than 125 million people in the United States now wear glasses or contact lenses.

The History of Optics

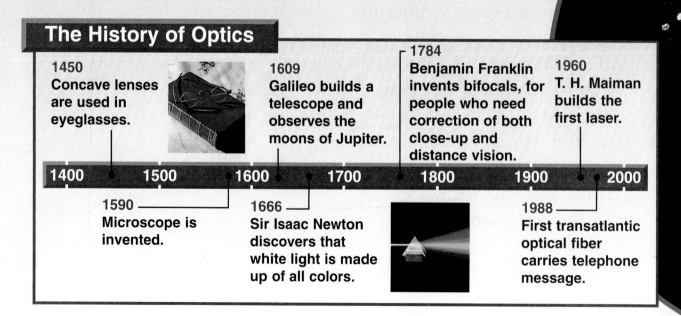

1450
Concave lenses are used in eyeglasses.

1590
Microscope is invented.

1609
Galileo builds a telescope and observes the moons of Jupiter.

1666
Sir Isaac Newton discovers that white light is made up of all colors.

1784
Benjamin Franklin invents bifocals, for people who need correction of both close-up and distance vision.

1960
T. H. Maiman builds the first laser.

1988
First transatlantic optical fiber carries telephone message.

1400   1500   1600   1700   1800   1900   2000

Lasers—Light in a Straight Line

If you've ever shone a flashlight into a dark room, you've seen a property of most light beams. The beams spread apart as they leave their source. Lasers turn a regular beam of light into a narrow, straight beam of bright light. Laser light is very focused and has only one color.

Laser light is used in many ways. Lasers are used to scan bar codes on products. Laser light has been bounced off the moon to accurately measure its distance from Earth.

◀ **Fiber Optics**

Physicians use lasers to do surgery. The most common use of lasers is in compact disc (CD) players. A laser beam cuts information onto the discs. The narrow beam allows a disc to hold more information than a tape. Lasers are then used to read and play back the recorded information. Besides music, entire encyclopedias have been put on CDs.

Telephones have long used electric current and copper wire to carry messages. Flashes of light can be used to send messages, too. Laser beams can carry many different messages along very thin glass fibers called optical fibers. Many fibers, each carrying a different message, can be squeezed into a single cable. Fiber-optic telephone lines are now used between many cities. Lines were laid across the Atlantic and Pacific Oceans in the late 1980s.

Fiber optics are also used in medicine to make surgery easier. Doctors can use the fibers to see inside the body while making only small cuts—or no cuts at all.

Think About It

1. How can lenses change an image?
2. What are two uses of optical fibers?

Lewis Howard Latimer
INVENTOR, ENGINEER

Every time you turn on an electric light, you can thank Lewis Latimer. His many inventions helped improve the first light bulb, which had been made by Thomas Edison. And if you've ever screwed a light bulb into a socket, you have used one of Latimer's inventions. He designed the threads of the socket. His model was made of wood, but we still use his idea.

Latimer was the youngest son of escaped slaves. He had to leave school when he was ten to earn money for the family. He never stopped learning, though. He taught himself mechanical drawing by watching the men in the office where he worked. They made detailed drawings of inventions for patent

applications. (Having a patent means the inventor "owns" the idea and the invention.) Latimer's office was near the office of Alexander Graham Bell, who invented the telephone. When Bell applied for a patent, he asked Latimer to make the drawing.

Later Latimer worked with the Edison Pioneers, a group of 80 inventors. He was the only African American in the group. He helped install lighting systems in New York, Philadelphia, Montreal, and even London.

Think About It

1. What do inventing and writing poetry have in common?
2. How is teaching yourself something, perhaps by watching others, different from learning in a classroom?

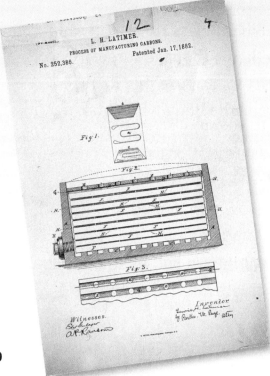

L. H. LATIMER.
PROCESS OF MANUFACTURING CARBONS.
No. 252,386.   Patented Jan. 17, 1882.

Colors

What colors are reflected off different colors of paper?

Materials

- glue
- strips of colored construction paper
- prism

Procedure

1. Glue strips of construction paper together in the order of the colors of the rainbow: red, orange, yellow, green, blue, and violet.

2. Use a prism to separate the colors in sunlight. Aim the colors from the prism at the different colors of construction paper.

3. Observe how the light from the prism is reflected by the different colors of construction paper.

Draw Conclusions

What colors from the prism are reflected from the green piece of construction paper? Explain.

Make a Periscope

How can you see around a corner?

Materials

- glue
- aluminum foil
- 2 index cards
- shoe box
- black construction paper

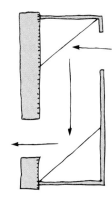

Procedure

1. Glue aluminum foil, shiny side out, to the index cards to make mirrors. Make the foil as smooth as possible.

2. Line the inside of the box with black paper. Cut out a hole in the bottom of the box, about 3 cm from one end. Cut a hole in the lid about 3 cm from one end.

3. Fold the ends of the aluminum foil mirrors to make tabs. Then glue the aluminum-foil mirrors to the inside of the box as shown.

4. Put the lid back on the box, and look through your periscope.

Draw Conclusions

How could you use a periscope to see around a corner?

Vocabulary Review

Use the terms below to complete the sentences 1 through 4. The page numbers in ( ) tell you where to look in the chapter if you need help.

reflection (F36)   **absorption** (F40)

refraction (F38)   **prism** (F44)

1. The bending of light is called ____.

2. A ____ breaks white light into colors.

3. The bouncing of light off objects is called ____.

4. Stopping light and holding it in is ____.

Connect Concepts

Follow the path of light as it travels. Use the terms in the Vocabulary Review to complete the concept map.

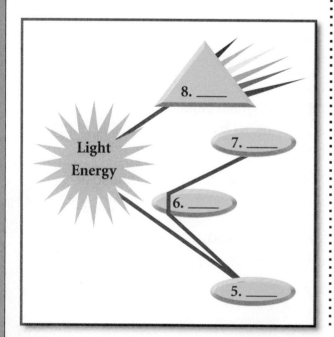

Check Understanding

Write the letter of the best choice.

9. Suppose you drop a penny into a shallow pool of water. You try to grab it but cannot seem to get your fingers in the right place. This happens because of —
   A reflection
   B absorption
   C refraction
   D light energy

10. Suppose you are standing at a pond. Your friend tries to sneak up on you, but you see him coming. You see him in the pond because of —
    F refraction
    G reflection
    H noise in the grass
    J absorption

11. White light is really —
    A all colors of light mixed
    B a mixture of yellow and white light
    C bright in the morning
    D a mixture of red and green light

12. Light travels —
   F through walls
   G around objects
   H in straight lines
   J in a curvy pattern

Critical Thinking

13. A skylight has water drops on it from a rainstorm. The sun comes out, and you see a rainbow on the wall. What is happening?

14. Describe how you could use a mirror to signal your friend in the house across the street.

15. You go to see a play. The light on the stage is yellow. You look up at the lights. They are red and green. Explain.

16. For art class, your teacher has you draw a bowl of fruit. The bowl contains a red apple, an orange, and a banana. After you have finished, your teacher puts a green spotlight on the fruit and asks you to draw it again. Why do you need to draw a new picture?

Process Skills Review

Write *True* or *False*. If the statement is false, correct it to make it true.

17. When you **observe** what is happening in an experiment, you use only your eyes.

18. Scientists sometimes draw pictures to explain their experiments.

Performance Assessment

Make a Model Prism

With a partner, use construction paper to make a large model of a prism breaking a ray of white light into its colors. Be sure to show the colors in the right order. Label each color. Make a hole in the model and add some string so it can be hung up in the classroom. You will need construction paper, glue, scissors, string, and a pencil.

Vocabulary Preview

force
motion
speed
gravity
weight
work
simple machine
lever
inclined plane

Forces and Motion

Do you ever wonder if you could throw a ball so fast that it wouldn't fall back to Earth? It takes a lot of energy to overcome gravity. Gravity is the force that pulls things back to Earth's surface. Rockets are the only human-made things that have been able to escape Earth's gravity and make it into space.

≡FAST FACT

You may never travel fast enough to escape Earth's gravity. But how fast can you go? A race car goes about 354 kilometers (220 mi) per hour. To go the 4715 kilometers (2,930 mi) between New York and San Francisco, it would take a race car $13\frac{1}{2}$ hours!

Time to Travel from New York to San Francisco		
Object	**Speed**	**Time**
8-year-old runner	19 kph	248 hours
Propeller plane	483 kph	$9\frac{3}{4}$ hours
Supersonic jet	2,173 kph	2 hours
Space shuttle	40,233 kph	7 minutes
Fast meteoroid	241,395 kph	1 minute
Light	299,330 km per second	0.016 second

How Do Forces Cause Motion?

In this lesson, you can . . .

INVESTIGATE how forces are measured.

LEARN ABOUT how forces act.

LINK to math, writing, social studies, and technology.

INVESTIGATE

Measuring Pushes and Pulls

Activity Purpose Suppose you pull an empty wagon down the street. Then a friend gets in the wagon, and you pull again. The second time, you have to pull a lot harder. Scientists use a tool called a spring scale to **measure** pulls. In this investigation you will use a spring scale to measure pulls.

Materials

- spring scale
- 2 pieces of string
- 2 wooden blocks

Activity Procedure

1 Work with a partner. One of you should hold one end of the spring scale while the other gently pulls on the hook. Read the number on the scale. The number is in newtons. Newtons measure pushes and pulls. **Record** the number. (Picture A)

◄ Pushes and pulls can make things move. A yo-yo moves up and down on a string as you drop it down and pull it back.

2 Tie one piece of string around one of the wooden blocks. Tie the other end of the string to the hook on the scale.

3 Begin to pull on the spring scale. Pull as hard as you can without making the block move. **Record** the number on the scale.

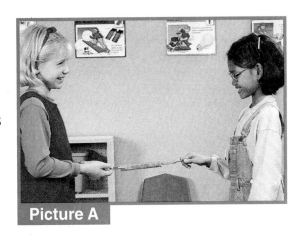

Picture A

4 Now pull hard enough to make the block move. **Record** the number on the scale. Carefully pull the block across the room. Watch the scale. Make sure the number doesn't change as you pull. (Picture B)

5 Repeat Step 4. This time, pull so that the number on the scale changes. **Record** your observations.

6 Tie the second wooden block to the first one. Repeat Steps 3 through 5. **Record** your observations.

Picture B

Draw Conclusions

1. How many newtons did it take to pull on the block without moving it? How many newtons did you use to move the block in Step 4?

2. How did you pull differently in Steps 4 and 5?

3. How did your results change when you added the second block?

4. **Scientists at Work** Scientists make charts to help them **interpret data**. Make a chart to organize the data you collected in this investigation.

Forces and Motion

VOCABULARY

force
motion
speed
gravity
weight

Forces

A **force** is a push or a pull. You push on a door. You pull on a wagon. Each push or pull is a force. There are many different forces. The force of the wind pushes sailboats and windmills. Forces in car engines can pull cars down the street.

In the investigation you pulled a wooden block. You used a spring scale to measure how hard you pulled. The spring scale measured the amount of force you used. The amount of force you have to use to move an object depends on its mass. The more mass something has, the more force you have to use to make it move. In the investigation you added a second block. How did that change the amount of force it took to move the blocks?

✔ **What is a force?**

◄ This boy is pushing the snowball. The push of the boy is a force.

This girl is pulling a wagon. The pull of the girl is a force. ►

Motion

Birds fly across the sky. Insects crawl across the ground. Leaves move back and forth in the breeze. The wheels on a bicycle go around in a circle. All these things are in motion. **Motion** is a change in position.

Every time you see something in motion, you know a force somewhere got it going. Every motion is started by a force. And every motion is stopped by a force. Once something is in motion, it will move until some force stops it.

✔ **What is motion?**

▲ Look at the picture of the river and the city. Each streak of light shows that an object is moving. What other motion can you see?

The speed of something tells how fast it is moving. Some things can move much faster than others. ▼

Snail
$\frac{5}{100}$ kilometers per hour (kmh) ($\frac{3}{100}$ miles per hour mph)

Person walking
5 kmh
(3 mph)

Person biking
60 kmh
(37 mph)

Race horse
70 kmh
(43 mph)

Motion

Suppose you want to move a soccer ball down the field. To start the ball moving, you have to apply a force. You give the ball a push with your foot. The ball starts moving. Now suppose you want to stop the ball. Again you have to apply a force. You have to put your foot up and push on the ball to stop it. If you don't stop the ball, it will keep rolling until some other force stops it.

Not all forces stop or start motion. Suppose you and a friend push on a box. You push one way. Your friend pushes the opposite way. The forces on the box are balanced, so the box doesn't move. But if another person starts pushing on your side, the

forces aren't balanced anymore. Now the box moves.

✔ **Why does a rolling ball stop rolling?**

When this clown rides her unicycle, the forces pushing on the left side are the same as the forces pushing on the right side. This keeps her from falling sideways. ▼

F60

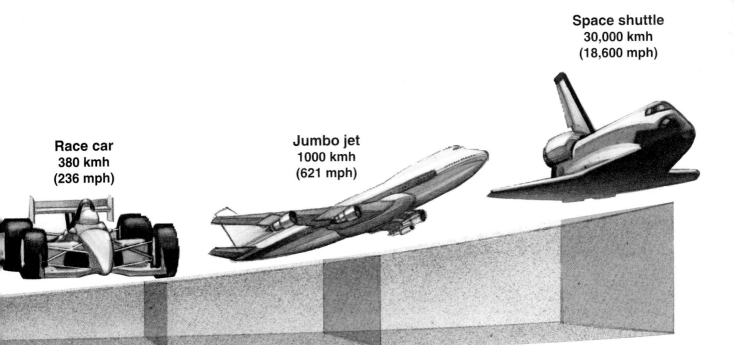

Race car
380 kmh
(236 mph)

Jumbo jet
1000 kmh
(621 mph)

Space shuttle
30,000 kmh
(18,600 mph)

Speed

Glaciers are large sheets of ice that move very slowly. They move less than 30 centimeters (about 1 ft) a day. A train can move much faster. A train can move more than 2000 kilometers (about 1,200 mi) in a day. The measure of how fast something moves over a certain distance is its **speed**. Something that moves a greater distance in the same amount of time has greater speed.

To find speed, you divide the distance you go by the time it takes you to get there. It's about 440 miles from Washington, D.C., to Boston. If you drive it by car in 8 hours, your speed is 440 miles divided by 8 hours, which is 55 miles per hour. An airplane can make the same trip in about 2 hours. The speed of the airplane is 880 miles divided by 2 hours, which is 440 miles per hour. Which has the greater speed, the car or the airplane?

✔ **What is speed?**

Glaciers move slowly. ▼

Gravity

There is a force that pulls us and everything around us down. That's why when you drop something, it falls down. The force is called gravity. **Gravity** (GRAV•ih•tee) is the force that pulls objects toward each other. All objects are acted on by gravity.

The more mass two things have, the more gravity pulls them toward each other. Earth has a very large mass, so the pull between objects and Earth is large. The moon has less mass than Earth. The pull between the moon and objects is about one-sixth of what it is on Earth.

Weight (WAYT) is a measure of the pull of gravity on an object. On Earth you have one weight. But if you were to go somewhere with a different amount of gravity, you would have a different weight. On the moon, for example, you would weigh one-sixth what you weigh on Earth.

✔ **What is weight?**

Gravity is the force that pulls objects toward Earth. On Earth, all things fall toward the ground. ▼

This astronaut is on the moon. His weight is $\frac{1}{6}$ what it is on Earth. If he weighs 180 lb on Earth, he weighs 30 lb on the moon. ▶

Summary

A force is a push or a pull. Motion is a change in position. All motion is caused by forces. Gravity is the force that pulls objects toward each other. Earth's gravity is so strong that it causes all things near Earth to fall toward Earth. Weight is a measure of gravity's pull on an object.

Review

1. What is force?
2. What happens if you start something moving and no force stops it?
3. What two things do you need to know to tell how fast something is moving?
4. **Critical Thinking** What would happen to a rock in a garden if no forces were ever applied to it?
5. **Test Prep** Which sentence is the best definition of *weight*?

   A Weight measures how big something is.

   B Weight measures how fast something can travel.

   C Weight measures the pull of gravity on an object.

   D Weight measures the height of an object.

LINKS

MATH LINK

Subtraction Story A cheetah can run as fast as 70 mph. A zebra can run 40 mph. How much faster can a cheetah run?

WRITING LINK

Expressive Writing—Poem
Write a short poem for your classmates, describing how the force of the wind moves the leaves on a plant.

SOCIAL STUDIES LINK

History of the Metric System
Find out where the metric system came from. Who invented it? What is the metric unit that measures force?

TECHNOLOGY LINK

To learn more about forces and motion, watch *Big Machine Summer* on the **Harcourt Science Newsroom Video.**

What Is Work?

In this lesson, you can . . .

INVESTIGATE
the scientific
definition of *work*.

LEARN ABOUT
what work is.

LINK to math,
writing, language arts,
and technology.

INVESTIGATE

Measuring Work

Activity Purpose Scientists say
that whenever you move an object, you do
work. When you make your bed, put books
onto shelves, or take out the trash, you do
work. In this investigation you will gather
information about work. Then you will
infer how work is related to force.

Materials

- mug
- spring scale
- meterstick
- shoe with laces
- hat
- stapler
- string

Activity Procedure

1 Make a table like the one shown.

2 Hook the mug onto the spring scale.
Work with a partner. One of you should
hold the meterstick on the table, with
the 1-cm mark toward the bottom. The
other person should rest the mug on the
table next to the meterstick. (Picture A)

◀ **A forklift can lift
heavy loads. It is
made to do work.**

Distance x Force = Work			
Object	Distance (centimeters)	Force (newtons)	Work (newton-centimeters)

3 Gently lift the scale, and pull the mug off the table. Keep the scale and the mug next to the meterstick. **Record** the height you lifted the mug. (Picture B)

4 Read the number of newtons on the spring scale. **Record** the number.

5 Multiply the number of newtons you used times the distance you moved the mug. This is the amount of work that was done. **Record** this number in your table under the heading *Work*.

6 Repeat Steps 2 through 5, using each of the other objects. You may need to use the string to attach the objects to the spring scale. Lift each object the same distance you lifted the mug.

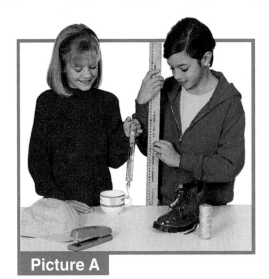
Picture A

Picture B

Draw Conclusions

1. Which object took the most force to lift? Which object took the least force to lift?

2. Describe the math you did in the chart.

3. In this investigation one of the measurements was kept the same. Which one was it?

4. **Scientists at Work** Scientists **infer** from data. Look at the data you collected. Infer how work is related to force.

Process Skill Tip

Scientists **infer** from data they collect when they **experiment**. After they make inferences, they continue experimenting to see if they are right.

Work

FIND OUT

- **what work is**
- **what force has to do with work**

VOCABULARY

work

Doing Work

For scientists the word *work* has a specific meaning. **Work** is the measure of force it takes to move an object a certain distance. In the investigation you did work when you lifted a mug off the table. Scientifically, you do work when you plant flowers, ride your bike, or help put away groceries. In each of these activities, you use a force to make something move. You don't do work when you push on a door and can't get it open. If you and a friend push against a box as hard as you can but the box doesn't move, you aren't doing any work.

This boy is pushing as hard as he can against a wall. He's not doing any work, though, because the wall doesn't move. ▼

▲ This time he pushes on a door. The door moves. Now he's doing work.

To figure out how much work is done, scientists multiply the force needed to move the object times the distance the object moves. If you use 3 newtons of force to pull a wagon 10 meters, you do 30 newton-meters of work.

✔ **What is work?**

Summary

In science, work is done when a force moves an object. If the object doesn't move, no work is done. To figure out the amount of work done, scientists multiply the force that is used times the distance the object moves.

Review

1. At the beginning of this lesson there is a picture of a forklift. Is it doing work? How do you know?

2. To a scientist, what is work?

3. How do scientists find the amount of work done?

4. **Critical Thinking** Describe a scene in which someone is using force but is not doing work.

5. **Test Prep** For which activity would you be doing work?

    A watching TV

    B pushing on a box that doesn't move

    C reading a page in a book

    D raking leaves

LINKS

MATH LINK

How Much Work? Marie used 3 newtons to move a chair 2 meters. Latifa used 2 newtons to lift her kitten 1 meter. Who did more work?

WRITING LINK

Informative Writing—How-To Write step-by-step instructions for your brother in first grade. Tell him how to do some kind of work. It might be how to hit a ball or how to make a bed. Be creative or funny.

LANGUAGE ARTS LINK

Many Meanings Write four sentences using the word *work*. Make two sentences with the scientific meaning and two sentences with the everyday meaning.

TECHNOLOGY LINK

Visit the Harcourt Learning Site for related links, activities, and resources.
www.harcourtschool.com

WELCOME TO
THE
LEARNING
SITE

What Are Simple Machines?

In this lesson, you can . . .

INVESTIGATE how simple machines help us do work.

LEARN ABOUT kinds of simple machines.

LINK to math, writing, social studies, and technology.

INVESTIGATE

Moving Up

Activity Purpose People often get help when they do work. Sometimes they ask friends to help. Sometimes they use machines to help. In this investigation you will **compare** two ways of doing work.

Materials

- string
- spring scale
- toy car
- chair
- meterstick
- wooden board (about 1 m long)
- masking tape

Activity Procedure

1. Using the string, attach the spring scale to the toy car. With the spring scale, slowly lift until the bottom of the car is at the same level as the seat of the chair. Read the spring scale, and **record** the number. (Picture A)

3. Use the meterstick to **measure** how high you lifted the car. **Record** this number, too.

◄ A nutcracker is an example of a simple machine called a lever. It makes work—in this case cracking nuts—seem easier to do.

3 Prop up one end of the board on the seat of the chair. Tape the board to the floor so the board does not move. (Picture B)

4 Now place the toy car at the bottom of the board. Slowly pull the car to the top of the board at a steady rate. As you pull, read the force on the spring scale. **Record** it.

5 **Measure** the distance you pulled the car up the board. **Record** it.

Picture A

Picture B

Draw Conclusions

1. Multiply the force you **measured** by the distance you moved the toy car. How much work did you do without the board? How much work did you do with the board?

2. **Compare** the force used without the board to the force used with the board. Which method used less force?

3. **Compare** the distance the car moved each time. Which method used less distance?

4. **Scientists at Work** Scientists often **compare** their data. Compare your answers to Question 1 to the answers of your classmates. What did you find out?

Investigate Further Repeat the activity using boards of several different lengths. **Record** the results. Did you use less force with longer boards or with shorter ones?

Process Skill Tip

Scientists look at the data they collect. They want to see what it means. Sometimes they have to **compare** two pieces of data. This means they look at the pieces of data to see how they are alike and how they are different.

Machines

Simple Machines

A **simple machine** is a tool that helps people do work. It doesn't change the amount of work. It just makes the work seem easier. Some simple machines reduce the amount of force you have to use. Other simple machines change the direction of the force you use. Some simple machines do both.

There are six simple machines. These machines are the lever, the pulley, the wheel and axle, the inclined plane, the wedge, and the screw. The board you used in the investigation was an inclined plane. When you pull something up an inclined plane, you use less force than you use to pick it straight up. But you have to move it a longer distance with the inclined plane. A pulley changes the direction of the force. As you pull down on one end of the rope, you lift up whatever is attached to the other end. Because of gravity, pulling down is easier than pulling up.

✔ **What is a simple machine?**

FIND OUT

- what a simple machine is
- names of simple machines

VOCABULARY

simple machine
lever
inclined plane

A **lever** (LEV·er) is a bar that moves on or around a fixed point. For a seesaw, the board you sit on is the bar. The fixed point is the stand. You can go up and down on the seesaw as it moves on the stand. ▶

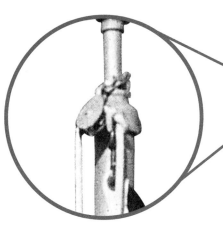

A pulley is used on a flagpole to raise and lower the flag. A pulley is a rope over a wheel. It changes the direction of the force you use. When you pull down on one end of the rope, the object attached to the other end moves up. ▶

◀ A wedge, such as this doorstop, is two inclined planes stuck back to back.

An **inclined plane** (IN·klynd PLAYN) is a flat surface set at an angle to another surface. It has many uses. Here it is used as a ramp for wheelchairs. ▼

▲ Look closely at a screw. It has an inclined plane winding around it. A screw is like an inclined plane wrapped around a pencil.

Scissors

Several simple machines put together make a compound machine. A pair of scissors is made up of two wedges and two levers. Look closely. Each blade is a wedge. The very narrow edges of the wedges meet and can cut through paper or fabric.

To make the blades meet every time, you need another machine. Each wedge moves around a bolt in the center. This bolt turns the wedges into levers.

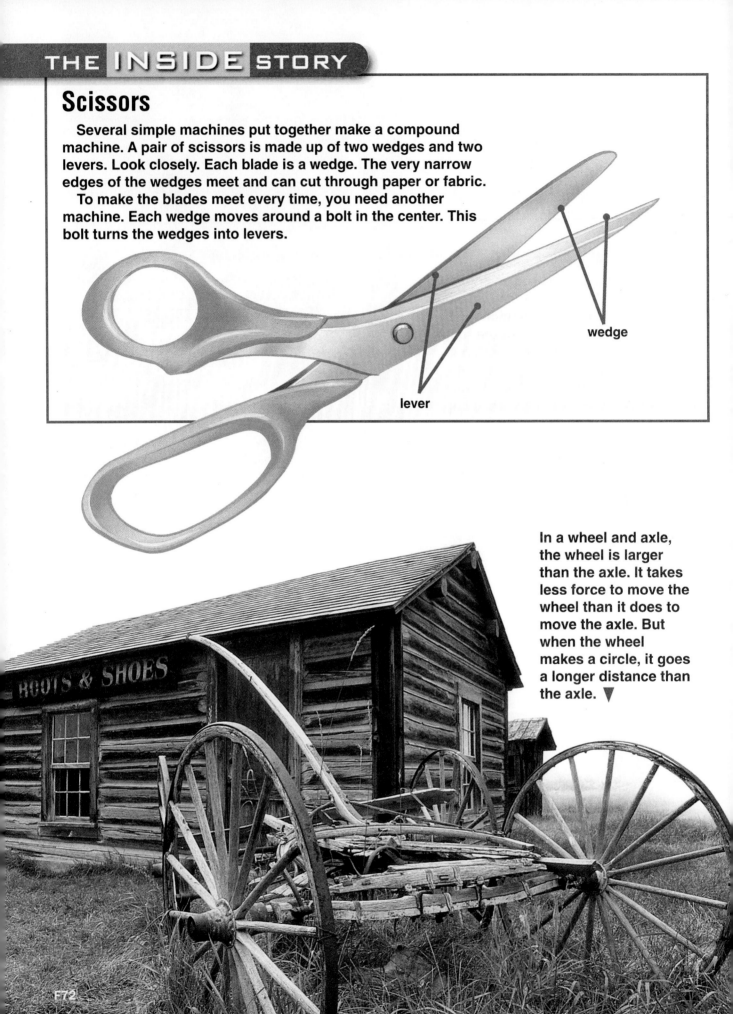

wedge

lever

In a wheel and axle, the wheel is larger than the axle. It takes less force to move the wheel than it does to move the axle. But when the wheel makes a circle, it goes a longer distance than the axle. ▼

Summary

Simple machines do not change the amount of work, but they make work seem easier. They reduce the amount of force you use or change the direction of the force. They can be used by themselves, or they can be parts of more complicated machines.

Review

1. Name two ways simple machines can make work seem easier.
2. What kind of simple machine is a ramp?
3. When you turn a wheel and axle, which moves farther, the wheel or the axle?
4. **Critical Thinking** A hand-cranked can opener contains a wheel and axle, a wedge, and two levers. Tell what part of a can opener each of these is.
5. **Test Prep** Which simple machine is a kind of inclined plane?
   A wheel and axle
   B pulley
   C screw
   D lever

LINKS

MATH LINK

A Word Problem Evan used 2 newtons of force to roll a rock 6 meters up a ramp. Janet attached the same rock to a pulley and used 6 newtons of force to raise it 2 meters. Who did more work?

WRITING LINK

Informative Writing— Description Write a paragraph describing a machine you have designed to make it easier to carry your books to school.

SOCIAL STUDIES LINK

Log Cabins How were log cabins built? In a group, talk about how the logs had to be cut and lifted into place. Decide which simple machines the pioneers may have used.

TECHNOLOGY LINK

Learn more about simple machines and how they work by investigating *Simple Machines* on **Harcourt Science Explorations CD-ROM.**

PROGRAMMABLE Toy Building Bricks

At first they were just plastic building blocks. Now you can use them to build simple machines that think on their own.

Why a Programmable Brick?

The official reason for making programmable bricks is that students can learn by using them. The real reason may be that it's fun.

It was surely fun to use plastic building blocks to make toy bridges and skyscrapers. But people wanted to do more with them. So scientists have made blocks that can be put together to make motors and other devices. These blocks are called bricks. It seems reasonable that the next thing they would do is make programmable bricks. Now you can make machines that you can program to move, act, and even think.

How Does It Work?

The programmable brick has places to plug in sensors that can "see" and "touch." There are also places to plug in motors and devices that make sounds, such as bells and whistles. The brick also has a transmitter like the one on your TV

remote control. So it can send signals to and receive signals from your computer.

The brick has a tiny processor that can handle 1,000 commands per second. You use your computer to program the brick. For example, you can program the brick to send a message to a motor to go backward or forward when the brick's sensor touches something.

What Can You Do with a Programmable Brick?

You can use the programmable brick to make a clever toy. Or you can use it to make an instrument that gathers data. For example, you can build an instrument that senses when you enter a room and turns on the light for you. Or you can build a robot similar to the Mars rover. You can program your rover to go around obstacles as it moves. Then you can send it to explore the house or back yard.

Two elementary school students used a programmable brick to make a device that took a picture every time a bird landed on their bird feeder. That way they could see what types of birds came to eat there.

The programmable brick was made to be used in robotics projects by students in elementary school,

junior high school, and high school. But that doesn't mean you can't have fun with it. With a little work and imagination, you might make a working model of a horse. Or you might build a robot that plays hockey!

Think About It

1. What other types of robot do you think you could make with a programmable brick?

2. How might a programmable brick be used to make a burglar alarm?

 WEB LINK:
For Science and Technology updates, visit the Harcourt Brace Internet site.
www.harcourtschool.com

Careers   Toy Designer

What They Do
Toy designers use their imagination to invent new toys that people will want to buy. They need to know how to  make toys that are safe, last a long time, and are easy to take care of.

Education and Training Some colleges offer courses in toy design. Toy design is a part of a degree in industrial design.

Christine Darden

ENGINEER

"Most of what you obtain in life will be because of your discipline. Discipline is perhaps more important than ability."

One of Christine Darden's childhood loves was fixing things. If her bicycle was broken, she tried to fix it herself. In high school, Darden knew that she wanted to study math.

After teaching math for several years, Darden went back to school herself. She began working as a mathematician at the National Aeronautics and Space Administration (NASA). There she became fascinated by the work that the engineers were doing.

Darden wanted to study engineering, but NASA didn't approve of the idea at first. It is not easy to switch fields of study. Finally, she persuaded NASA, and she returned to school where she earned a degree in engineering.

Darden is now working with others to make a plane that flies faster than the speed of sound but doesn't cause a sonic boom. Because of the noise they make, planes that fast are banned in the United States.

Darden was given the 1992 Women in Science and Engineering (WISE) Award. Every year WISE honors three women who have done important work in science and engineering.

Think About It

1. Which do you think is more important—finding a job easily or doing the work you love to do?
2. How did Christine Darden show her determination?

Movement from Air

How can you move a Ping Pong ball with your breath?

Materials
- Ping Pong ball
- 3 straws
- colorful tape

Procedure
1. Work in groups of three. Each person needs a straw. Clear a space on the floor. With tape, mark a course on the floor. Make sure the course has some twists and turns in it.

2. Put the Ping Pong ball at the beginning of the course. With your partners, blow on the ball to make it follow the path.

Draw Conclusions
From which angle did you blow to make the ball move farthest? Which made the ball move farther, blowing softly or blowing hard? Which made the ball move farther, blowing steadily or blowing in short puffs?

Measuring Weight

How can you make your own spring scale?

Materials
- paper cup
- a large, heavy rubber band
- 2 rulers
- string
- tape
- objects

Procedure
1. With your pencil, punch two small holes on each side of the paper cup. Place the rubber band as shown. Thread a piece of string through each pair of holes in the cup and tie the cup to the rubber band.

2. Hold the ruler straight, and have your partner use another ruler to measure the length of the rubber band from the top of the cup to the ruler.

3. Compare the weights of different objects by putting them in the cup and measuring the length of the rubber band.

Draw Conclusions
How is the length of the rubber band related to the weight of the objects put in the cup?

Chapter ③ Review and Test Preparation

Vocabulary Review

Choose a term below to match each definition. The page numbers in ( ) tell you where to look in the chapter if you need help.

force (F58) **simple machine** (F70)

motion (F59)

speed (F61) **lever** (F70)

weight (F62) **inclined**

gravity (F62) **plane** (F71)

work (F66)

1. How fast an object moves from one position to another

2. A flat surface set at an angle to another surface

3. A change in position

4. A force that pulls all objects toward each other

5. A tool that helps people do work

6. A bar that moves on or around a fixed point

7. A push or a pull

8. A measure of the pull of gravity on an object

9. Measure of a force moving an object

Connect Concepts

Study the diagram. Then use terms from the chapter to complete the sentences.

10. Your push on the truck is a ____.

11. The length of the ramp is the ____ the truck must move.

12. The distance the truck moves divided by the time it takes is the truck's ____.

13. The simple machine the truck is moving up is an ____.

14. The force to push the truck times the distance the truck moves is the ____ done to push the truck up the ramp.

Check Understanding

Write the letter of the best choice.

15. A rock sat on the edge of a cliff. One day the rock fell off. The force that caused this motion was probably —

   A energy **C** work

   B a pulley **D** gravity

16. Workers at an airport put suitcases on an airplane. They use a moving ramp. What kind of simple machine is it?

  F an inclined plane   **H** a wedge

  G a screw   **J** a pulley

17. A jet flies at 550 mph. This is a measure of —

  A height   **C** fuel

  B speed   **D** weight

18. Alice uses 2 newtons of force to push her little sister's stroller 5 meters. How much work does she do?

  F 2 newton-meters

  G 5 newton-meters

  H 7 newton-meters

  J 10 newton-meters

Critical Thinking

19. Suppose you use a screwdriver to pry the top off a can. Explain which kinds of simple machines you use and how they work for this job.

20. Anthony moved dirt all day with a wheelbarrow. He moved it 20 meters from the vegetable garden to the flower garden. Max sat in his room and studied science. Who would scientists say did more work? Why?

Process Skills Review

21. A red car and a blue car both travel to a town 60 miles away. The red car gets there first. What can you **infer** about the speeds of the two cars?

22. You tried four different ways to lift a bag of sand 1 meter. With the pulley, you used 5 newtons of force. You used 3 newtons when you dragged the bag up the inclined plane. When you pushed the bag up with the lever, you used 2 newtons of force. And when you picked it up by yourself, you used 5 newtons of force. Make a table to help you **interpret the data. Compare** the methods. Tell which used the most force and which used the least force.

Performance Assessment

Motion and Work

  A. Using a book on your desk, show a force causing motion. Identify the force.

  B. Using the book again, show a force that doesn't cause motion. Was work done in **A**, in **B**, or in both?

Unit Project Wrap Up

Here are some ideas for ways to wrap up your unit project.

Write a Program

Write a program for your shadow show. Include a short summary of the story. Tell about the different kinds of forces and energy used in the show.

Invent a Toy

Invent a toy that has moveable parts. Your toy should have at least one part that is a simple machine. Draw a diagram of your invention, and label the simple machine.

Make a Booklet

Fold and staple paper into a booklet. Write the name of a simple machine on the top of each page. Look through magazines for pictures of items that have simple machine parts. Cut out the pictures and glue them on the right pages of your booklet.

Investigate Further

How could you make your project better? What other questions do you have about energy and forces? Plan ways to find answers to your questions. Use the Science Handbook on pages R2-R9 for help.

References

Science Handbook

Planning an Investigation

When scientists observe something they want to study, they use scientific inquiry to plan and conduct their study. They use science process skills as tools to help them gather, organize, analyze, and present their information. This plan will help you work like a scientist.

Step 1—Observe and ask questions.

Which food does my hamster eat the most of?

- Use your senses to make observations.
- Record a question you would like to answer.

Step 2—Make a hypothesis.

My hypothesis: My hamster will eat more sunflower seeds than any other food.

- Choose one possible answer, or hypothesis, to your question.
- Write your hypothesis in a complete sentence.
- Think about what investigation you can do to test your hypothesis.

Step 3—Plan your test.

I'll give my hamster equal amounts of three kinds of foods, then observe what she eats.

- Write down the steps you will follow to do your test. Decide how to conduct a fair test by controlling variables.
- Decide what equipment you will need.
- Decide how you will gather and record your data.

Step 4 — Conduct your test.

I'll repeat this experiment for four days. I'll meaure how much food is left each time.

- Follow the steps you wrote.
- Observe and measure carefully.
- Record everything that happens.
- Organize your data so that you can study it carefully.

Step 5—Draw conclusions and share results.

My hypothesis was correct. She ate more sunflower seeds than the other kinds of foods.

- Analyze the data you gathered.
- Make charts, graphs, or tables to show your data.
- Write a conclusion. Describe the evidence you used to determine whether your test supported your hypothesis.
- Decide whether your hypothesis was correct.

Investigate Further

I wonder if there are other foods she will eat . . .

Using Science Tools

Using a Hand Lens

1. Hold the hand lens about 12 centimeters (5 in.) from your eye.
2. Bring the object toward you until it comes into focus.

Using a Thermometer

1. Place the thermometer in the liquid. Never stir the liquid with the thermometer. Don't touch the thermometer any more than you need to. If you are measuring the temperature of the air, make sure that the thermometer is not in line with a direct light source.
2. Move so that your eyes are even with the liquid in the thermometer.
3. If you are measuring a material that is not being heated or cooled, wait about two minutes for the reading to become stable, or stay the same. Find the scale line that meets the top of the liquid in the thermometer, and read the temperature.
4. If the material you are measuring is being heated or cooled, you will not be able to wait before taking your measurements. Measure as quickly as you can.

Caring for and Using a Microscope

Caring for a Microscope

- Carry a microscope with two hands.
- Never touch any of the lenses of a microscope with your fingers.

Using a Microscope

1. Raise the eyepiece as far as you can using the coarse-adjustment knob. Place your slide on the stage.

2. Start by using the lowest power. The lowest-power lens is usually the shortest. Place the lens in the lowest position it can go to without touching the slide.

3. Look through the eyepiece, and begin adjusting it upward with the coarse-adjustment knob. When the slide is close to being in focus, use the fine-adjustment knob.

4. When you want to use a higher-power lens, first focus the slide under low power. Then, watching carefully to make sure that the lens will not hit the slide, turn the higher-power lens into place. Use only the fine-adjustment knob when looking through the higher-power lens.

You may use a Brock microscope. This sturdy microscope has only one lens.

1. Place the object to be viewed on the stage.

2. Look through the eyepiece, and raise the tube until the object comes into focus.

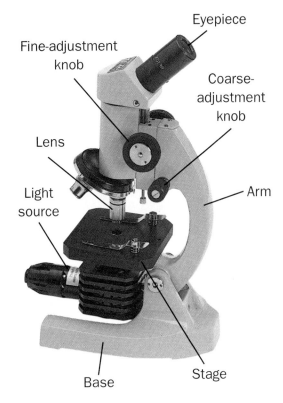

A Light Microscope

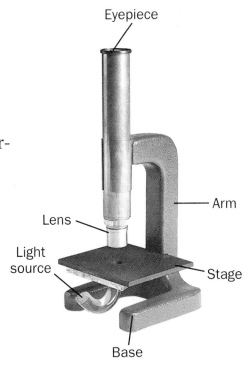

A Brock Microscope

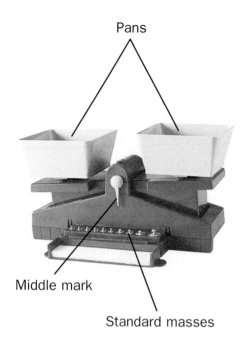

Pans

Middle mark

Standard masses

Using a Balance

1. Look at the pointer on the base to make sure the empty pans are balanced. Place the object you wish to measure in the left-hand pan.

2. Add the standard masses to the other pan. As you add masses, you should see the pointer move. When the pointer is at the middle mark, the pans are balanced.

3. Add the numbers on the masses you used. The total is the mass in grams of the object you measured.

Using a Spring Scale

Measuring an Object at Rest

1. Hook the spring scale to the object.

2. Lift the scale and object with a smooth motion. Do not jerk them upward.

3. Wait until any motion of the spring comes to a stop. Then read the number of newtons from the scale.

Measuring an Object in Motion

1. With the object resting on a table, hook the spring scale to it.

2. Pull the object smoothly across the table. Do not jerk the object.

3. As you pull, read the number of newtons you are using to pull the object.

Measuring Liquids

1. Pour the liquid you want to measure into a measuring container. Put your measuring container on a flat surface, with the measuring scale facing you.

2. Look at the liquid through the container. Move so that your eyes are even with the surface of the liquid in the container.

3. To read the volume of the liquid, find the scale line that is even with the surface of the liquid.

4. If the surface of the liquid is not exactly even with a line, estimate the volume of the liquid. Decide which line the liquid is closer to, and use that number.

Beaker   **Graduate**

Using a Ruler or Meterstick

1. Place the zero mark or end of the ruler or meterstick next to one end of the distance or object you want to measure.

2. On the ruler or meterstick, find the place next to the other end of the distance or object.

3. Look at the scale on the ruler or meterstick. This will show the distance or the length of the object.

Using a Timing Device

1. Reset the stopwatch to zero.

2. When you are ready to begin timing, press *Start*.

3. As soon as you are ready to stop timing, press *Stop*.

4. The numbers on the dial or display show how many minutes, seconds, and parts of seconds have passed.

Using a Computer

Writing Reports

To write a report with a computer, use a word processing software program. After you are in the program, type your report. By using certain keys and the mouse, you can control how the words look, move words, delete or add words and copy them, check your spelling, and print your report.

Save your work to the desktop or hard disk of the computer, or to a floppy disk. You can go back to your saved work later if you want to revise it.

There are many reasons for revising your work. You may find new information to add or mistakes you want to correct. You may want to change the way you report your information because of who will read it.

Computers make revising easy. You delete what you don't want, add the new parts, and then save. You can also save different versions of your work.

For a science lab report, it is important to show the same kinds of information each time. With a computer, you can make a general format for a lab report, save the format, and then use it again and again.

Making Graphs and Charts

You can make a graph or chart with most word processing software programs. You can also use special software programs such as Data ToolKit or Graph Links. With Graph Links you can make pictographs and circle, bar, line, and double-line graphs.

First, decide what kind of graph or chart will best communicate your data. Sometimes it's easiest to do this by sketching your ideas on paper. Then you can decide what format and categories you need for your graph or chart. Choose that format for the program. Then type your information. Most software programs include a tutor that gives you step-by-step directions for making a graph or chart.

Doing Research

Computers can help you find current information from all over the world through the Internet. The Internet connects thousands of computer sites that have been set up by schools, libraries, museums, and many other organizations.

Get permission from an adult before you log on to the Internet. Find out the rules for Internet use at school or at home. Then log on and go to a search engine, which will help you find what you need. Type in keywords, words that tell the subject of your search. If you get too much information that isn't exactly about the topic,

make your keywords more specific. When you find the information you need, save it or print it.

Harcourt Science tells you about many Internet sites related to what you are studying. To find out about these sites, called Web sites, look for Technology Links in the lessons in this book.

If you need to contact other people to help in your research, you can use e-mail. Log into your e-mail program, type the address of the person you want to reach, type your message, and send it. Be sure to have adult permission before sending or receiving e-mail.

Another way to use a computer for research is to access CD-ROMs. These are discs that look like music CDs. CD-ROMs can hold huge amounts of data, including words, still pictures, audio, and video. Encyclopedias, dictionaries, almanacs, and other sources of information are available on CD-ROMs. These computer discs are valuable resources for your research.

Visit the Multimedia Science Glossary to see illustrations of these words and to hear them pronounced.

www.harcourtschool.com/scienceglossary

Glossary

This Glossary contains important science words and their definitions. Each word is respelled as it would be in a dictionary. When you see the ′ mark after a syllable, pronounce that syllable with more force than the other syllables. The page number at the end of the definition tells where to find the word in your book. The boldfaced letters in the examples in the Pronunciation Key that follows show how these letters are pronounced in the respellings after each glossary word.

PRONUNCIATION KEY

a	**a**dd, m**a**p	m	**m**ove, see**m**	u	**u**p, d**o**ne
ā	**a**ce, r**a**te	n	**n**ice, ti**n**	û(r)	b**ur**n, t**er**m
â(r)	**c**are, **air**	ng	ri**ng**, so**ng**	yoo	**f**use, **few**
ä	p**a**lm, f**a**ther	o	**o**dd, h**o**t	v	**v**ain, e**v**e
b	**b**at, ru**b**	ō	**o**pen, s**o**	w	**w**in, a**w**ay
ch	**ch**eck, cat**ch**	ô	**or**der, j**aw**	y	**y**et, **y**earn
d	**d**og, ro**d**	oi	**oi**l, b**oy**	z	**z**est, mu**s**e
e	**e**nd, p**e**t	ou	p**ou**t, n**ow**	zh	vi**s**ion, plea**s**ure
ē	**e**qual, tr**ee**	o͝o	t**oo**k, f**u**ll	ə	the schwa, an
f	**f**it, hal**f**	o͞o	p**oo**l, f**oo**d		unstressed vowel
g	**g**o, lo**g**	p	**p**it, sto**p**		representing the
h	**h**ope, **h**ate	r	**r**un, poo**r**		sound spelled
i	**i**t, g**i**ve	s	**s**ee, pa**ss**		*a* in a*bove*
ī	**i**ce, wr**i**te	sh	**s**ure, ru**sh**		*e* in *sicken*
j	**j**oy, le**dg**e	t	**t**alk, si**t**		*i* in *possible*
k	**c**ool, ta**k**e	th	**th**in, bo**th**		*o* in *melon*
l	**l**ook, ru**l**e	t͟h	**th**is, ba**th**e		*u* in *circus*

Other symbols:

• separates words into syllables

′ indicates heavier stress on a syllable

′ indicates light stress on a syllable

A

absorption [ab·sôrp′shən] The stopping of light **(F40)**

amphibian [am·fib′ē·ən] An animal that begins life in the water and moves onto land as an adult **(A50)**

anemometer [an′ə·mom′ə·tər] An instrument that measures wind speed **(D40)**

asteroid [as′tər·oid] A chunk of rock that orbits the sun **(D64)**

atmosphere [at′məs·fir′] The air that surrounds Earth **(D30)**

atom [at′əm] The basic building block of matter **(E16)**

axis [ak′sis] An imaginary line that goes through the North Pole and the South Pole of Earth **(D68)**

B

barrier island [bar′ē·ər ī′lənd] A landform; a thin island along a coast **(C35)**

bird [bûrd] An animal that has feathers, two legs, and wings **(A45)**

C

canyon [kan′yən] A landform; a deep valley with very steep sides **(C35)**

chemical change [kem′i·kəl chānj′] A change that forms different kinds of matter **(E46)**

chlorophyll [klôr′ə·fil′] The substance that gives plants their green color; it helps a plant use energy from the sun to make food **(A20)**

clay [klā] A type of soil made up of very small grains; it holds water well **(C69)**

coastal forest [kōs′təl fôr′ist] A thick forest with tall trees that gets a lot of rain and does not get very warm or cold **(B15)**

comet [kom′it] A large ball of ice and dust that orbits the sun **(D64)**

community [kə·myoo′nə·tē] All the populations of organisms that live in an ecosystem **(B7)**

condensation [kon′dən·sā′shən] The changing of a gas into a liquid **(D17)**

conductor [kən·duk′tər] A material in which thermal energy moves easily **(F15)**

coniferous forest [kō·nif′ər·əs fôr′ist] A forest in which most of the trees are conifers (cone-bearing) and stay green all year **(B16)**

conservation [kon′ser·vā′shən] The saving of resources by using them carefully **(C76)**

constellation [kon′stə•lā′shən] A group of stars that form a pattern **(D84)**

consumer [kən•sōōm′ər] A living thing that eats other living things as food **(B43)**

contour plowing [kon′tōōr plou′ing] A type of plowing for growing crops; creates rows of crops around the sides of a hill instead of up and down **(C76)**

core [kôr] The center of the Earth **(C8)**

crust [krust] The solid outside layer of the Earth **(C8)**

deciduous forest [dē•sij′ōō•əs fôr′ist] A forest in which most of the trees lose and regrow their leaves each year **(B13)**

decomposer [dē′kəm•pōz′er] A living thing that breaks down dead organisms for food **(B44)**

desert [dez′ərt] An ecosystem where there is very little rain **(B20)**

earthquake [ûrth′kwāk′] The shaking of Earth's surface caused by movement of the crust and mantle **(C48)**

ecosystem [ek′ō•sis′təm] The living and non-living things in an environment **(B7)**

energy [en′ər•jē] The ability to cause change **(F6)**

energy pyramid [en′ər•jē pir′ə•mid] A diagram that shows that the amount of useable energy in an ecosystem is less for each higher animal in the food chain **(B50)**

environment [in•vī′rən•mənt] The things, both living and nonliving, that surround a living thing **(B6)**

erosion [i•rō′zhən] The movement of weathered rock and soil **(C42)**

estuary [es′chōō•er′•ē] A place where fresh water from a river mixes with salt water from the ocean **(D12)**

evaporation [ē•vap′ə•rā′shən] The process by which a liquid changes into a gas **(D17, E18)**

fish [fish] An animal that lives its whole life in water and breathes with gills **(A52)**

flood [flud] A large amount of water that covers normally dry land **(C50)**

food chain [fōōd′ chān′] The path of food from one living thing to another **(B48)**

food web [fōōd′ web′] A model that shows how food chains overlap **(B54)**

force [fôrs] A push or a pull **(F58)**

forest [fôr′ist] An area in which the main plants are trees **(B12)**

fossil [fos′əl] Something that has lasted from a living thing that died long ago **(C20)**

fresh water [fresh′ wôt′ər] Water that has very little salt in it **(B26)**

front [frunt] A place where two air masses of different temperatures meet **(D37)**

gas [gas] A form of matter that does not have a definite shape or a definite volume **(E12)**

germinate [jûr′mə•nāt′] When a new plant breaks out of the seed **(A13)**

gills [gilz] A body part found in fish and young amphibians that takes in oxygen from the water **(A51)**

glacier [glā′shər] A huge sheet of ice **(C44)**

gravity [grav′i•tē] The force that pulls objects toward each other **(F62)**

groundwater [ground′wôt′ər] A form of fresh water that is found under Earth's surface **(D8)**

habitat [hab′ə•tat′] The place where a population lives in an ecosystem **(B7)**

heat [hēt] The movement of thermal energy from one place to another **(F8)**

humus [hyo͞o′məs] The part of the soil made up of decayed parts of once-living things **(C62)**

igneous rock [ig′nē•əs rok′] A rock that was once melted rock but has cooled and hardened **(C12)**

inclined plane [in•klīnd′ plān′] A simple machine made of a flat surface set at an angle to another surface **(F71)**

inexhaustible resource [in′eg•zôs′tə•bəl rē′sôrs] A resource such as air or water that can be used over and over and can't be used up **(C94)**

inherit [in•her′it] To receive traits from parents **(A38)**

insulator [in′sə•lāt′ər] A material in which thermal energy does not move easily **(F15)**

interact [in′tər•akt′] When plants and animals affect one another or the environment to meet their needs **(B42)**

landform [land′fôrm′] A natural shape or feature of Earth's surface **(C34)**

leaf [lēf] A plant part that grows out of the stem; it takes in the air and light that a plant needs **(A7)**

lever [lev′ər] A bar that moves on or around a fixed point **(F70)**

liquid [lik′wid] A form of matter that has volume that stays the same, but can change its shape **(E12)**

loam [lōm] A type of topsoil that is rich in minerals and has lots of humus **(C70)**

lunar eclipse [loo′nər i•klips′] The hiding of the moon when it passes through the Earth's shadow **(D78)**

mammal [mam′əl] An animal that has fur or hair and is fed milk from its mother's body **(A42)**

mantle [man′təl] The middle layer of the Earth **(C8)**

mass [mas] The amount of matter in an object **(E24)**

matter [mat′ər] Anything that takes up space **(E6)**

metamorphic rock [met′ə•môr′fik rok′] A rock that has been changed by heat and pressure **(C12)**

mineral [min′ər•əl] An object that is solid, is formed in nature, and has never been alive **(C6)**

mixture [miks′chər] A substance that contains two or more different types of matter **(E41)**

motion [mō′shən] A change in position **(F59)**

mountain [moun′tən] A landform; a place on Earth's surface that is much higher than the land around it **(C35)**

nonrenewable resource [non′ri•noo′ə•bəl rē′sôrs] A resource, such as coal or oil, that will be used up someday **(C96)**

orbit [ôr′bit] The path an object takes as it moves around another object in space **(D58)**

phases [fāz•əz] The different shapes the moon seems to have in the sky when observed from Earth **(D76)**

photosynthesis [fōt′ō•sin′thə•sis] The food-making process of plants **(A20)**

physical change [fiz′i•kəl chānj] A change to matter in which no new kinds of matter are formed **(E40)**

physical property [fiz′i•kəl prop′ər•tē] Anything you can observe about an object by using your senses **(E6)**

plain [plān] A landform; a flat area on Earth's surface **(C35)**

planet [plan′it] A large body of rock or gas that orbits the sun **(D58)**

plateau [pla•tō′] A landform; a flat area higher than the land around it **(C35)**

population [pop′yōō•lā′shən] A group of the same kind of living thing that all live in one place at the same time **(B7)**

precipitation [prē•sip′ə•tā′shən] The water that falls to Earth as rain, snow, sleet, or hail **(D18)**

predator [pred′ə•tər] An animal that hunts another animal for food **(B54)**

prey [prā] An animal that is hunted by a predator **(B54)**

prism [priz′əm] A solid, transparent object that bends light into colors **(F44)**

producer [prə•dōōs′ər] A living thing that makes its own food **(B43)**

recycle [rē•sī′kəl] To reuse a resource to make something new **(C100)**

reflection [ri•flek′shən] The bouncing of light off an object **(F36)**

refraction [ri•frak′shən] The bending of light when it moves from one kind of matter to another **(F38)**

renewable resource [ri•nōō′ə•bəl rē′sôrs] A resource that can be replaced in a human lifetime **(C94)**

reptile [rep′tīl] A land animal that has dry skin covered by scales **(A55)**

resource [rē′sôrs] A material that is found in nature and that is used by living things **(C88)**

revolution [rev′ə•lōō′shən] The movement of one object around another object **(D68)**

rock [rok] A solid made of minerals **(C8)**

rock cycle [rok′ sī′kəl] The process in which one type of rock changes into another type of rock **(C14)**

root [rōōt] The part of a plant that holds the plant in the ground and takes in water and minerals from the soil **(A7)**

rotation [rō•tā′shən] The spinning of an object on its axis **(D68)**

salt water [sôlt′ wôt′ər] Water that has a lot of salt in it **(B26)**

scales [skālz] The small, thin, flat plates that help protect the bodies of fish and reptiles **(A52)**

sedimentary rock [sed′ə•men′tər•ē rok′] A rock formed from material that has settled into layers and been squeezed until it hardens into rock **(C12)**

seed [sēd] The first stage in the growth of many plants **(A12)**

seedling [sēd′ling] A young plant **(A13)**

simple machine [sim′pəl mə•shēn′] A tool that helps people do work **(F70)**

soil [soil] The loose material in which plants can grow in the upper layer of Earth **(C62)**

solar eclipse [sō′lər i•klips′] The hiding of the sun that occurs when the moon passes between the sun and Earth **(D80)**

solar system [sō′lər sis′təm] The sun and the objects that orbit around it **(D58)**

solid [sol′id] A form of matter that takes up a specific amount of space and has a definite shape **(E11)**

solution [sə•lōō′shən] A mixture in which the particles of two different kinds of matter mix together evenly **(E42)**

speed [spēd] The measure of how fast something moves over a certain distance **(F61)**

star [stär] A hot ball of glowing gases, like our sun **(D84)**

stem [stem] A plant part that connects the roots with the leaves of a plant and supports the plant above ground; it carries water from the roots to other parts of the plant **(A7)**

strip cropping [strip′ krop′ing] A type of planting that uses strips of thick grass or clover between strips of crops **(C76)**

telescope [tel′ə•skōp′] An instrument used to see faraway objects **(D88)**

temperature [tem′pər•ə•chər] The measure of how hot or cold something is **(D36)**

thermal energy [thûr′məl en′ər•jē] The energy that moves the particles in matter **(F7)**

thermometer [thûr•mom′ə•tər] A tool used to measure temperature **(F20)**

topsoil [top′soil′] The top layer of soil made up of the smallest grains and the most humus **(C63)**

trait [trāt] A body feature that an animal inherits; it can also be some things that an animal does **(A38)**

tropical rain forest [trop′i•kəl rān′fôr′ist] A hot, wet forest where the trees grow very tall and their leaves stay green all year **(B14)**

valley [val′ē] A landform; a lowland area between higher lands, such as mountains **(C35)**

volcano [vol•kā′nō] An opening in Earth's surface from which lava flows **(C49)**

volume [vol′yo͞om] The amount of space that matter takes up **(E22)**

water cycle [wôt′ər sī′kəl] The movement of water from Earth's surface into the air and back to the surface again **(D19)**

weather [weth′ər] The happenings in the atmosphere at a certain time **(D32)**

weather map [weth′ər map′] A map that shows weather data for a large area **(D46)**

weathering [weth′ər•ing] The process by which rock is worn down and broken apart **(C40)**

weight [wāt] The measure of the pull of gravity on an object **(F62)**

wind [wind] The movement of air **(D40)**

work [wûrk] The measure of force that it takes to move an object a certain distance **(F66)**

Photography Credits - Page placement key: (t) top, (c) center, (b) bottom, (l) left, (r) right, (bg) background, (i) inset

Cover Background, Charles Krebs/Tony Stone Images; Inset, Jody Dole.

Table of Contents - iv (bg) Thomas Brase/Tony Stone Images; (i) Denis Valentine/The Stock Market; v (bg) Derek Redfeam/The Image Bank; (i) George E. Stewart/Dembinsky Photo Association; vi (bg) Richard Price/FPG International; (i) Martin Land/Science Photo Library/Photo Researchers; vii (bg) Pal Hermansen/Tony Stone Images; (i) Earth Imaging/Tony Stone Images; viii (bg) Steve Barnett/Liaison International; (i) StockFood America/Lieberman; ix (bg) Simon Fraser/Science Photo Library/Photo Researchers; (i) Nance Trueworthy/Liaison International.

Unit A - A1 (bg) Thomas Brase/Tony Stone Images; (i) Denis Valentine/The Stock Market; A2-A3 (bg) Joe McDonald/Bruce Coleman; A3 (i) Marilyn Kazmers/Deminsky Photo Associates; A4 Ed Young/AGStock USA; A6 (l) Anthony Edgeworth/The Stock Market; (r) Chris Vincent/The Stock Market; A6-A7 (bg) Barbara Gerlach/Dembinsky Photo Associates; A7 (c) Wendy W. Cortesi; A8 (t) Runk/Schoenberger/Grant Heilman Photography; (c) Runk/Schoenberger/Grant Heilman Photography; (bl) Renee Lynn/Photo Researchers; (br) Dr. E.R. Degginger/Color-Pic; A9 Runk/Schoenberger/Grant Heilman Photography; A10 Runk/Schoenberger/Grant Heilman Photography; A12 (l) Bonnie Sue/Grant Heilman/Photo Researchers; (li) Klaus Paysan/Peter Arnold, Inc.; (r) Runk/Schoenberger/Grant Heilman Photography; (ri) Runk/Schoenberger/Grant Heilman Photography; A13 (tr) Ed Young/AgStock USA; (b) Dr. E. R. Degginger/Color-Pic; A14 (tr) Richard Shiell/Dembinsky Photo Associates; (bl) Robert Carr/Bruce Coleman, Inc.; (br) Scott Sinklier/AGStock USA; A16 (t) Thomas D. Mangelsen/Peter Arnold, Inc.; (c) E.R. Degginger/Natural Selection Stock Photography; (bl) Randall B. Henne/Dembinsky Photo Associates; (br) Stan Osolinski/Dembinsky Photo Associates; (I) Scott Camazine/Photo Researchers; A17 William Harlow/Photo Researchers; A18 Christi Carter/Grant Heilman Photography; A20 Runk/Schoenberger/Grant Heilman Photography; A22 (t) DiMaggio/Kalish/The Stock Market; (cl) Jan-Peter Lahall/Peter Arnold, Inc.; (br) Holt Studios/Nigel Cattlin/Photo Researchers; A23 Robert Carr/Bruce Coleman; A24 Richard Shiell; A25 J. Sapinsky/The Stock Market; A26 (tr) Corbis; A30-A31 (bkgd) Art Wolfe/Tony Stone Images; A31 (cr) Astrid & Hanns Frieder Michler/Science Photo Library/Photo Researchers; A32 (bl) Rosemary Calvert/Tony Stone Images; (1) Ralph A. Reinhold/Animals Animals; (2) Johnny Johnson/ Tony Stone Images; (3) Mike Severns/ Tom Stack & Associates; (4) Fred Whitehead/Animals Animals; (5) Art Wolfe/ Tony Stone Images; (6) J.C. Stevenson/ Animals Animals; A34-A35 Doug Perrine/Innerspace Visions; A35 (t) Ronald Hellstrom/Bruce Coleman, Inc.; (br) Stan Osolinski/Tony Stone Images; A36 (tr) Mike Severns/Tony Stone Images; (lc) Kevin Schafer Photography; (bl) Marilyn Kazmers/Peter Arnold, Inc.; (br) Keren Su/Tony Stone Images; A38 (t) Rudie Kuiter/Innerspace Visions; (c) Fred Bruemmer/Peter Arnold, Inc.; (b) Art Wolfe/Tony Stone Images; A39 Phil A. Dotson/Photo Researchers; A40 Brian Stablyk/Tony Stone Images; A43 (t) Paul Metzger/Photo Researchers; (b) Frans Lanting/Minden Pictures; A44 (t) Stephen Dalton/Photo Researchers; (c) Tom McHugh/Photo Researchers; (cr) Evelyn Gallardo/Peter Arnold, Inc.; (cl) The Photo Library-Sydney/Gary Lewis/Photo Researchers; (br) Francois Gohier/Photo Researchers; A45 (t) Theo Allofs/Tony Stone Images; (blue jay) Wayne Lankinen/Bruce Coleman, Inc.; (macaw) M. Mastrorillo/The Stock Market; (emperor penquin) Kjell B. Sandved/Photo Researchers; (ostrich) Leonard Lee Rue III/Photo Researchers; (bee humming bird) Robert A. Tyrrell Photography; (peacock) Tom McHugh/Photo Researchers; A46 (t) Manfred Danegger/Tony Stone Images; (cl) John Cancalosi/Peter Arnold, Inc.; (b) Bill Ivy/Tony Stone Images; (br) Stan Osolinski/The Stock Market; A48 (c) O.S.F./ Animals Animals; (b) Tim Davis/Tony Stone Images; A50 (tl) Nuridsany et Perennou/Photo Researchers; (r) E.R. Degginger/Photo Researchers; (bl) Joseph T. Collins/Photo Researchers; A52 (t) David M. Schleser/Nature's Images; (c) Andrea & Antonella Ferrari/Innerspace Visions; A52-A53 Kelvin Aitken/Peter Arnold, Inc.; A53 (t) Zig Leszczynski/Animals Animals; (c) Kelvin Aitken/Peter Arnold, Inc.; (br) Tom McHugh/Steinhart Aquarium/Photo Researchers; A54(t) Kim Taylor/Bruce Coleman, Inc.; (c) Fred Bavendam/Minden Pictures; (b) Fred Bavendam/Minden Pictures; A55 (t) Zig Leszczynski/Animals Animals; (b) Suzanne L. Collins & Joseph T. Collins/Photo Researchers; (bli) Dwight R. Kuhn; A56 (t) Jany Sauvanet/Photo Researchers; (c) G.E. Schmida/Fritz/Bruce Coleman, Inc.; A56-A57 (b) Tom & Pat Leeson/Photo Researchers; A57 Schafer & Hill/Tony Stone Images; A58 (t) Tom Brakefield/Bruce Coleman, Inc.; (c) Dr. E. R. Degginger/Color-Pic; (b) Michael Holford; A59 Emory Kristof/National Geographic Image Collection; A60 (tr) Bertha G. Gomez; (bl) Michael Fogden/bruce Coleman, Inc.

Unit B - B1 (bg) Derek Redfeam/The Image Bank; (i) George E. Stewart/Dembinsky Photo Association; B2-B3 (bg) Sven Linoblad/Photo Researchers; B2 (i) Wayne P. Armstrong; B4 Hans Pfletschinger/Peter Arnold, Inc.; B6 (l) Dwight R. Kuhn; (r) Michael Durham/ENP Images; B7 Frank Krahmer/Bruce Coleman, Inc.; B8 (tl) Jeff and Alexa Henry/Peter Arnold, Inc.; (t) Jeff and Alexa Henry/Peter Arnold, Inc.; (b) Christoph Burki/Tony Stone Images; B10 Kennan Ward/The Stock Market; B13 (all) James P. Jackson/Photo Researchers; B14 Zefa Germany/The Stock Market; B15 Janis Burger/Bruce Coleman Inc.; B16 (r) Michael Quinton/Minden Pictures; B16-B17 (b)Grant Heilman Photography; B18 J.C. Carton/Bruce Coleman, Inc.; B20 (l) Wolfgang Kaehler Photography; (r) James Randklev/Tony Stone Images; B21 Dr. E.R. Degginger/Color-Pic; B22 (l) Paul Chesley/Tony Stone Images; (r) Jeff Foott/Bruce Coleman, Inc.; (i) Jen & Des Bartlett/Bruce Coleman, Inc.; B23 Lee Rentz/Bruce Coleman, Inc.; B24 Leo De Wys Inc.; B27 (li) R.N. Mariscal/Bruce Coleman, Inc.; (b) Dr. E.R. Degginger/Color-Pic; (ri) Naitar E. Harvey, APSA/National Audubon Society/Photo Researchers; B28 Flip Nicklin/Minden Pictures; B29 (t) Norbert Wu/Peter Arnold, Inc.; (b) Norbert Wu/Peter Arnold, Inc.; B30 (t) Gary Meszaros/Bruce Coleman, Inc.; (bli) Stevan Stefanovic/Okapia/Photo Researchers; (bci) Dwight R. Kuhn; (bri) Phil Degginger/Color-Pic; B30-B31 (b) Jeff Greenberg/Photo Researchers; B32 (b) Courtesy of Jane Weaver/Parie Project/L. A. Gilillard Elementary; (ti) Globe-NASA/ Goddard Scientific Visualization Studio; B33 Derke/O'Hara/Tony Stone Images; B34 (tr) The Marjorie N. Boyer Trust; (bl) Anthony Mercieca/ Photo Researchers; B38-B39 Luiz C. Marigo/Peter Arnold, Inc.; B39 (br) Roland Seitre/Peter Arnold, Inc.; B40 (l) Roy Morsch/The Stock Market; (c) Norbert Wu/Tony Stone Images; (r) Rosemary Calvert/Tony Stone Images;
B41 (bl) Stan Osolinski/The Stock Market; (c) R. Kopfle/KOPFL/Bruce Coleman; (br) Michael Durham/ENP Images; B42 (bg) Hans Reinhard/Bruce Coleman, Inc.; (li) Dwight R. Kuhn; (ri) Dr. Paul A. Zahl/Photo Researchers; B43 (t) Wolfgang Kaehler Photography; (b) Rob Hadlow/Bruce Coleman, Inc.; B44 (t) Stephen Dalton/Photo Researchers; (c) Andrew Syred/Science Photo Library/Photo Researchers; (b) Stephen Krasemann/Tony Stone Images; B46 Laurie Campbell/Tony Stone Images; B48 Dwight R. Kuhn; B49 (t) Paul E. Taylor/Photo Researchers; (c) Holt Studios/Photo Researchers; (b) Breck P. Kent/Animals Animals; B51 Mitsuaki Iwago/Minden Pictures; B52 Erwin and Peggy Bauer/Bruce Coleman, Inc.; B54-B55 Michael Durham/ENP Images; B57 Jane Burton/Bruce Coleman; B58 (cl) LASCAUX Caves II, France/Explorer, Paris/Superstock; (c) Fred Bruemmer/Peter Arnold; (bc) Tom Brakefield/Bruce Coleman; B60 (tl) Leah Edelstein-Keshet/University of British Columbia; (bl) Fred McConnaughey/Photo Researchers.

Unit C Other - C1(bg) Richard Price/FPG International; (i) Martin Land/Science Photo Library/Photo Researchers; C2-C3 (bg) E. R. Degginger; C2 (bc) A. J. Copley/Visuals Unlimited; C3 (ri) Paul Chesley/Tony Stone Images; C4 (b) The Natural History Museum, London; C6 (tl), (ct), (cb) Dr. E. R. Degginger/Color-Pic; (r) E. R. Degginger/Bruce Coleman, Inc.; (bl) Mark A. Schneider/Dembinsky Photo Associates; C6-C7 (b) Chromosohm/Joe Sohm/Photo Researchers; C7 (tr) Blair Seitz/Photo Researchers; (tri), (bli) Dr. E. R. Degginger/Color-Pic; C8 (t) Barry Runk/Grant Heilman Photography; (ct) Dr. E. R. Degginger/Color-Pic; (b) Dr. E. R. Degginger/Color-Pic; (cb) Barry L. Runk/Grant Heilman Photography; C8-9 (t) Robert Pettit/Dembinsky Photo Associates; C10 Tom Bean/Tom & Susan Bean, Inc.; C12 Jim Steinberg/Photo Researchers; C12-C13 (bg) G. Brad Lewis/Photo Resource Hawaii; C14 (l), (tr) Dr. E. R. Degginger/Color-Pic; C14 (br) Aaron Haupt/Photo Researchers; C15 (tl) Robert Pettit/Dembinsky Photo Associates; (tr) Charles R. Belinky/Photo Researchers; (bl), (bc), (br) Dr. E. R. Degginger/Color-Pic; C16 (t) Roger Du Buisson/The Stock Market; (c) Jay Mallin Photos; C16-C17 (b) Ed Wheeler/The Stock Market; C18 Stephen Wilkes/The Image Bank; C21 (t) William E. Ferguson; (bl) Kerry T. Givens/Bruce Coleman, Inc.; (br) Joy Spurr/Bruce Coleman, Inc.; C22 (t) AP Photo/Dennis Cook; (b) M. Timothy O'Keefe/Bruce Coleman, Inc.; C24 (t) Francois Gohier/Photo Researchers; (b) The National History Museum/London; C25 Stan Osolinski; C26 (tr) Jean Miele/Lamont-Doherty Earth Observatory of Columbia University; C30-C31 (bg) John Warden/Tony Stone Images; C31 (bl) Harold Naideau/The Stock Market; C32 G. Alan Nelson/Dembinsky Photo Associates; C33 (b) Superstock; C34-35 (b) Darrell Gulin/Dembinsky Photo Associates; C35 (t) Michael Hubrich/Dembinsky Photo Associates; (c) Mark E. Gibson; C36 (t) Breck P. Kent/Earth Scenes; C36-37 (b) Paraskevas Photography; C38 Mark E. Gibson; C40 (bl) Dr. E.R. Degginger/Color-Pic; C40 (bc) Mark A. Schneider/Dembinsky Photo Associates; C40-C41 (br-b) Rod Planck/Dembinsky Photo Associates; C41 (t) Michael Hubrich/Dembinsky Photo Associates; (c) John Gerlach/Dembinsky Photo Associates; C42 (t) Georg Gerster/Photo Researchers; (c) NASA Photo/Grant Heilman Photography; C42-C43 (b) C.C. Lockwood/Earth Scenes; C43 (t) Mark E. Gibson; C46 Ken Sakamoto/Black Star; C48 (l) David Parker/SPL/Photo Researchers; (l) AP/Wide World Photos; C49 (l) AP/Wide World Photos; (r) AP Photo/Wide World Photos; (bl) Will & Deni McIntyre/Photo Researchers; C50-C51 AP/Wide World Photos; C52 (l) George Hall/Woodfin Camp & Associates; (i) Laura Riley/Bruce Coleman; C53 J. Aronovsky/Zuma Images/The Stock Market; C54 (tr) Courtesy of Scott Rowland; (bl) Dennis Oda/Tony Stone Images; C58-C59 (bg) Lynn M. Stone/Bruce Coleman, Inc.; C59 (br) NASA; C60 Ann Duncan/Tom Stack & Associates; C63 (all) Bruce Coleman, Inc.; C66 Grant Heilman/Grant Heilman Photography; C68-C69 (b) Gary Irving/Panoramic Images; C68 (I) Barry L. Runk/Grant Heilman Photography; C69 (li), (ri) Barry L. Runk/Grant Heilman Photography; C70-C71 (b) Larry Lefever/Grant Heilman Photography; C72 Andy Sacks/Tony Stone Images; C74 USDA - Soil Conservation Service; C74-C75 (b) Dr. E.R. Degginger/Color-Pic; C75 (t) James D. Nations/D. Donne Bryant; (tli) Gunter Ziseler/Peter Arnold, Inc.; (tri) S.AM./Wolfgang Kaehler Photography; (bli) Walter H. Hodge/Peter Arnold, Inc.; (bri) Jim Steinberg/Photo Researchers; C76 (t) Thomas Hovland from Grant Heilman Photography; (b) B.W. Hoffmann/AGStock USA; C78 (b) Randall B. Henne/Dembinsky Photo Associates; (I) Russ Munn/AgStock USA; C79 Bruce Hands/Tony Stone Images; C80 (tr) Courtesy of Diana Wall, Colorado State University; (bl) Oliver Mickes/Ottawa/Photo Researchers; C84-C85 (bg) Kirby, Richar OSF/Earth Scenes; C86 Bob Daemmrich/Bob Daemmrich Photography, Inc.; C88 (l) Peter Correz/Tony Stone Images; (c) Mark E. Gibson; C88-C89(b) Bill Lea/Dembinsky Photo Associates; C90 (t) Chris Rogers/Rainbow/PNI; (c) Yoav Levy/Phototake/PNI; C91 Rob Badger Photography; C92 (bl) Christie's Images, London/Superstock; (br) Jeff Greenberg / Photo Researchers; (bc) Joyce Photographics / Photo Researchers; (tr) Mary Ann Kulla/ The Stock Market; C93 (bl) Alan L. Detrick / Photo Researchers; (bc) Archive Photos; (br) David Barnes /The Stock Market; (tr) Gary Retherford/ Photo Researchers; C94-C95 Jeff Greenberg/Visuals Unlimited; C95 (tl) Wolfgang Fischer/Peter Arnold, Inc.; (tr) Wolfgang Fischer/Peter Arnold, Inc.; (c) Craig Hammell/The Stock Market; C96 (t) Barbara Gerlach/Dembinsky Photo Associates; (b) Brownie Harris/The Stock Market; C97 Chris Rogers/The Stock Market; C98 Michael A. Keller/The Stock Market; C100-C101 (b) Ray Pfortner/Peter Arnold, Inc.; C103 William E. Ferguson; C104-C105 Blaine Harrington III/The Stock Market; C106 James King-Holmes/Science Photo Library/ Photo Researchers; C107 (bl) Wellman Fibers Industry; (r) Gabe Palmer/The Stock Market; C108 (tr) Susan Sterner/HRW; (bl) Kristin Finnegan/Tony Stone Images.

Unit D Other - D1(bg) Pal Hermansen/Tony Stone Images; (i) Earth Imaging/Tony Stone Images; D2-D3 (bg) Zefa Germany/The Stock Market; D3 (tr) Michael A. Keller/The Stock Market; (br) Steven Needham/Envision; D4 J. Shaw/Bruce Coleman, Inc.; D5 (b) NASA; D6 (l) Yu Somov; S. Korytnikov/Sovfoto/Eastfoto/PNI; (r) Christopher Arend/Alaska Stock Images/PNI; D7 Grant Heilman Photography; D8-D9 Dr. Eckart Pott/Bruce Coleman, Inc.; D10 N.R. Rowan/The Image Works; D12-D13 Mike Price/Bruce Coleman, Inc.; D13 David Job/Tony Stone Images; D14 John Beatty/Tony Stone Images; D17 (l) Grant Heilman Photography; (r) Darrell Gulin/Tony Stone Images; D20 NASA; D21 Ben Osborne/Tony Stone Images; D22 (tr) Polytechnic State University; (b) NASA; D26-D27 Andrea Booher/Tony Stone Images; D27 NASA/Science Photo Library/Photo Researchers; D28 Joe Towers/The Stock Market; D30 Rich Iwasaki/Tony Stone Images; D32 (t) Peter Arnold; (c) Warren Faidley/International Stock Photography; (b) Stephen Simpson/FPG International; D33 E.R. Degginger/Bruce Coleman, Inc.; D34 Ray Pfortner/Peter Arnold, Inc.; D37 (t) Ralph H. Wetmore, II/Tony Stone Images; (b) Joe McDonald/Earth Scenes; D38 (c) Tom Bean; (b) Adam Jones/Photo Researchers; D42 Warren Faidley/International Stock Photography; D44 (l) Superstock; (r) David Ducros/Science Photo Library/Photo Researchers; D45 (both) © 1998 AccuWeather; D45 New Scientist Magazine; D49 Dwayne Newton/PhotoEdit; D50 (tr) Courtesy June Bacon-Bercey; (bl) David R. Frazier/Photo Researchers; D54-D55 David Hardy/Science Photo Library/Photo Researchers; D55 (tc) European Space Agency/Science Photo Library/Photo Researchers; D60 (t) U.S. Geological Survey/Science Photo Library/Photo Researchers; D60 (b) NASA; D60, D61, D62 (bg) Jerry Schad/Photo Researchers; D61 (t) National Oceanic and Atmospheric Administration; D61 (b) David Crisp and the WFPC2 Science Team (Jet Propulsion Laboratory/California Institute of Technology); D62 (t), (b) NASA; D63 (t) Erich Karkoschka (University of Arizona Lunar & Planetary Lab) and NASA; (c) NASA; (b) Nasa/Science Source/Photo Researchers; D64 J. Spurr/Bruce Coleman, Inc.; D65 Royer, Ronald/Science Photo Library/Photo Researchers; D66 Renee Lynn/Photo Researchers; D69 (l) Dr. E. R. Degginger/Color-Pic; (r) Dr. E. R. Degginger/Color-Pic; D72 (t) Joseph Nettis/Photo Researchers; (b) John Elk III/Bruce Coleman, Inc.; D74 NASA; D77 (all) Telegraph Colour Library/FPG International; D78-D79 Margaret Miller/Photo Researchers; D79 (t) Pekka Parviainen/Science Photo Library/Photo Researchers; (b) George East/Science Photo Library/Photo Researchers; D80 Dr. Fred Espenak/Science Photo Library/Photo Researchers; D82 The Granger Collection, New York; D88 Merritt Vincent/PhotoEdit; D90 (l) Rob Talbot/ Tony Stone Images; (r) Stephen Graham/Dembinsky Photo Associates; D91 NASA; D92 (both) NASA.

Unit E Other - E1(bg) Steve Barnett/Liaison International; (i) StockFood America/Lieberman; E2-E3 Chris Noble/Tony Stone Images; E3 Kent & Donna Dannen; E6 John Michael/International Stock Photography; E7 (r) R. Van Nostrand/Photo Researchers; E8 (tr) Mike Timo/Tony Stone Images; E9 (t) Daniel J. Cox/Tony Stone Images; (ti) Goknar/Vogue/Superstock; E10 (br) John Michael/International Stock Photography; (bc) William Cornett/Image Excellence Photography; E11 (tr) Lee Foster/FPG International; (br) Paul Silverman/Fundamental Photographs; E12 (t) William Johnson/Stock, Boston; E12-E13 (b) Robert Finken/Photo Researchers; E14 S.J. Krasemann/Peter Arnold, Inc.; E16 (ri) Dr. E. R. Degginger/Color-Pic; E17 (l) Charles D. Winters/Photo Researchers; (br) Spencer Gran/PhotoEdit; E18-E19 Peter French/Pacific Stock; E20 J. Sebo/Zoo Atlanta; E25 (cr) Robert Pearcy/Animals Animals; E25 (cl) Ron Kimball Photography; E26 (tr) Jim Harrison/Stock, Boston; E32 (tr) Corbis; (b) Alfred Pasieka/Science Photo Library/Photo Researchers; E36-E37 (bg)Dr. Dennis Kunkel/Phototake; E38 Robert Ginn/PhotoEdit; E42 (tl), (tr) Dr. E. R. Degginger/Color-Pic; (bl) Tom Pantages; (br) Tom Pantages; E44 Chip Clark; E46 (l) Tom Pantages; (r), (c) Tom Pantages; E47 (bl) John Lund/Tony Stone Images; E50 John Gaudio; E51 Michael Newman/Photo Edit; E52 (tr) Los Alamos National Laboratory/Photo Researchers; (bl) US Army White Sands Missile Range.

Unit F - F1 (bg) Simon Fraser/Science Photo Library/Photo Researchers; (i) Nance Trueworthy/Liaison International; F2 (bg) April Riehm; (tr) M. W. Black/Bruce Coleman, Inc.; F6 (bl) Mary Kate Denny/PhotoEdit; (br) Gary A. Conner/PhotoEdit; F7 (bl) Camerique, Inc./The Picture Cube; (br) Stephen Saks/The Picture Cube; F8 (cr) Pat Field/Bruce Coleman, Inc.; (bl) Mark E. Gibson; F9 (l) Dr. E.R. Degginger/Color-Pic; (r)Ryan and Beyer/Allstock/PNI; F10 (b) John Running/Stock Boston; (cl) Joseph Nettis/Photo Researchers; F11 (t) Jeff Schultz/Alaska Stock Images; F16 (t) Buck Ennis/Stock, Boston; (b) J.C. Carton/Bruce Coleman, Inc.; F21 (tl) Michael Holford Photographs; (br) Spencer Grant/PhotoEdit; F25 Marco Cristofori/ The Stock Market; F26 (tr) Corbis; (bl) Shaun Egan/Tony Stone Images; F30-F31 (bg) Jerry Lodriguss/Photo Researchers; F31 (br) Picture PerfectUSA; F32 (b) James M. Mejuto Photography; F34 (bl) Mark E. Gibson; (br) Bob Daemmrich/Stock, Boston; F36 (b) Myrleen Ferguson/PhotoEdit; F37 (b), (tl) Jan Butchofsky/Dave G. Houser; F39 (tr), (c) Richard Megna/Fundamental Photographs; F42 (b) Randy Duchaine/The Stock Market; F44 (b) Tom Skrivan/The Stock Market; F45 (tr) David Woodfall/Tony Stone Images; F47 (t) Roy Morsh/The Stock Market; F48 (l) Ed Eckstein for The Franklin Institute Science Museum; (r) Peter Angelo Simon/The Stock Market; F48-F49 Paul Silverman/ Fundamental Photography; F54-F55 (bg) Superstock; (br) David Madison/Bruce Coleman, Inc.; F58 (cl) John Running/Stock, Boston; (b) David Young-Wolff/PhotoEdit; F59 H. Mark Weidman; F60 (t) D & I McDonald/The Picture Cube; F62 (b) Nasa/The Stock Market; (r) Richard Megna/Fundamental Photographs; F64 (b) Edith G. Haun/Stock, Boston; F70 (b) Amy C. Etra/PhotoEdit; F71 (t) Tony Freeman/PhotoEdit; F72 (b) Dave G. Houser; F74 Webb Chappell; F75 John Lei/Omni-Photo Communications; F76 (tr) NASA/Langley Research Center; (bl) Valder/Tormey/International Stock.

Health Handbook - R15 Palm Beach Post; R19 (tr) Andrew Speilman/Phototake; (c) Martha McBride/Unicorn; (br) Larry West/FPG International; R21 Superstock; R26 (c) Index Stock; R27 (tl) Renne Lynn/ Tony Stone Images; (tr) David Young-Wolff/PhotoEdit.

All Other photographs by Harcourt photographers listed below, © Harcourt:
Weronica Ankarorn, Bartlett Digital Photography, Victoria Bowen, Eric Camdem, Digital Imaging Group, Charles Hodges, Ken Karp, Ken Kinzie, Ed McDonald, Sheri O'Neal, Terry Sinclair.

Illustration Credits - Craig Austin A53; Graham Austin B56; John Butler A42; Rick Courtney A20, A51, B14, B55, C22, C40, C41, C62, C64; Mike Dammer A27, A61, B35, B61, C27, C55, C81, C109, D23, D51, D93, E35, E53, F27, F51, F77; Dennis Davidson D58; John Edwards D9, D17, D18, D31, D37, D39, D40, D59, D64, D68, D69, D70, D71, D76, D77, D78, D80, E17, D10, F45, F60; Wendy Griswold-Smith A37; Lisa Frasier F78; Geosystems C36, C42, C44, C103, D12, D46; Wayne Hovice B28; Tom Powers C8, C13, C20, C34, C50, C102; John Rice D16, B7, B21, B50; Ruttle D36, D7; Rosie Saunders A15; Shough C90, F22, F46, F71, F72.